THE ISDN CON

Hewlett-Packard Professional Books

Atchison	Object-Oriented Test & Measurement Software Development in C++
Blinn	Portable Shell Programming: An Extensive Collection of Bourne Shell Examples
Blommers	Practical Planning for Network Growth
Costa	Planning and Designing High Speed Networks Using 100VG-AnyLAN, Second Edition
Crane	A Simplified Approach to Image Processing: Classical and Modern Techniques
Fernandez	Configuring the Common Desktop Environment
Fristrup	USENET: Netnews for Everyone
Fristrup	The Essential Web Surfer Survival Guide
Grady	Practical Software Metrics for Project Management and Process Improvement
Grosvenor, Ichiro, O'Brien	Mainframe Downsizing to Upsize Your Business: IT-Preneuring
Gunn	A Guide to NetWare® for UNIX®
Helsel	Graphical Programming: A Tutorial for HP VEE
Helsel	Visual Programming with HP-VEE
Kane	PA-RISC 2.0 Architecture
Knouse	Practical DCE Programming
Lee	The ISDN Consultant: A Stress-Free Guide to High-Speed Communications
Lewis	The Art & Science of Smalltalk
Madell, Parsons, Abegg	Developing and Localizing International Software
Malan, Letsinger, Coleman	Object-Oriented Development at Work: Fusion In the Real World
McFarland	X Windows on the World: Developing Internationalized Software with X, Motif®, and CDE
McMinds/Whitty	Writing Your Own OSF/Motif Widgets
Phaal	LAN Traffic Management
Poniatowski	The HP-UX System Administrator's "How To" Book
Poniatowski	HP-UX 10.x System Administration "How To" Book
Poniatowski	Learning the HP-UX Operating System
Thomas	Cable Television Proof-of-Performance: A Practical Guide to Cable TV Compliance Measurements Using a Spectrum Analyzer.
Weygant	Clusters for High Availability: A Primer of HP-UX Solutions
Witte	Electronic Test Instruments

THE ISDN CONSULTANT:

A Stress-Free Guide to High-Speed Communications

Robert E. Lee

Hewlett-Packard Company

Prentice Hall PTR
Upper Saddle River, New Jersey 07458

Acquisitions editor: *Karen Gettman*
Editorial Assistant: *Barbara Alfieri*
Production supervision: *Mary Sudul*
Copyeditor: *Barbara Danziger*
Cover design: *Talar Agasyan*
Cover design director: *Jerry Votta*
Manufacturing manager: *Alexis Heyd*t
Manager, Hewlett-Packard Press: *Patricia Pekary*

© 1997 by Hewlett-Packard Company

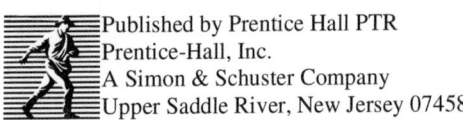
Published by Prentice Hall PTR
Prentice-Hall, Inc.
A Simon & Schuster Company
Upper Saddle River, New Jersey 07458

The publisher offers discounts on this book when ordered in bulk quantities.
For more information, contact Corporate Sales Department,
Prentice Hall PTR, One Lake Street, Upper Saddle River, NJ 07458.
Phone: 800-382-3419; Fax: 201-236-7141; E-mail: corpsales@prenhall.com

All product names mentioned herein are the trademarks of their respective owners.

All rights reserved. No part of this book may be
reproduced, in any form or by any means,
without permission in writing from the publisher.

Printed in the United States of America

10 9 8 7 6 5 4 3 2 1

ISBN 0-13-259052-2

Prentice-Hall International (UK) Limited, *London*
Prentice-Hall of Australia Pty. Limited, *Sydney*
Prentice-Hall Canada Inc., *Toronto*
Prentice-Hall Hispanoamericana, S.A., *Mexico*
Prentice-Hall of India Private Limited, *New Delhi*
Prentice-Hall of Japan, Inc., *Tokyo*
Simon & Schuster Asia Pte. Ltd., *Singapore*
Editora Prentice-Hall do Brasil, Ltda., *Rio de Janeiro*

For Vicky, Erich, and Christopher
The reason I try so hard.

Table of Contents

Foreword xv

Preface xvii

    Audience xviii
    Organization xx
    Conventions xxi
    Acknowledgments xxiii

Chapter 1 — Why All the Hoopla? 1

    Applications — That's Why! 1
    What is ISDN? 2
        Evolution from the Telephone 3
        Going Digital 3
        Enter ISDN 5
    Variation on a theme — the standards 6
    Applications — the process 7

Rising Expectations 7
Designing Applications 8
State of ISDN 9
Emerging Competitors 12
    Cable Modems 12
    Digital Subscriber Line 16
    Conclusions 17
What's Next 18

Chapter 2 — Explaining the Technology 19

nB+D — What's Inside the ISDN Pipe 20
    Phone Company Facilities 21
    BRI — Basic Rate Interface 24
        D Channel 25
        B Channel 25
    PRI — Primary Rate Interface 26
    Messaging 30
Current and Emerging ISDN Standards 32
    Standards Bodies 32
    National ISDN and Services 34
        National ISDN-1 34
        National ISDN-2 35
        National ISDN-3 37
    Protocols 38
Broadband ISDN — Bandwidth on Demand 41
    SMDS — Switched Multimegabit Data Service 41
    ATM — Asynchronous Transfer Mode 42
What's Next 44

Chapter 3 — Explaining the Hardware 45

Wiring 45
    Wire Categories 46
    Wire Types 46
    Number of Wires and Connectors 48
    Wire Closets 49

TABLE OF CONTENTS

    Reusing Wire 51
Network Terminators 52
    Interface Points 52
    Network Termination Devices 54
The ISDN equipment — Phones, Computers, and Video 55
    Telephone Systems 55
        PBX or Key Systems 55
        Inverse Multiplexor..56
        Telephones 56
    Computer Systems 57
        Terminal Adapters 57
        LAN Routers 59
        Network Hubs 60
    Videoconferencing Systems 60
        Conference Room Systems 60
        Desktop Systems..61
        Picture Phones...61
        Multimedia Conferencing 62
What's Next? 62

Chapter 4 — Ordering the Lines 63

Language of the carriers..63
    Basic Terms 64
        DN..64
        SPID..64
        TEI 66
    Voice Features 67
    Data Features 69
    Capability Packages 70
    Order Process 71
Crossing Carrier Boundaries..75
Tariffs — Understanding the Service Offerings 76
    Tariff Influences 76
    Regulatory Process 78
    Tariff Components 79
        Goods and Services..79

       Pricing and Commitments 80
       Final Comment 81
  What's Next 81

Chapter 5 — Designing Applications 83

  The Business Problem Defined 84
       Who Defines the Problem? 84
       What Does a Problem Definition Look Like? 86
  Solution Development 88
       Idea Generation 88
       Narrow the Solutions 89
       Determining the Capabilities of Your Existing Systems 92
           Wiring 93
           Existing Equipment 93
           People Skills 94
       Identifying the Missing Links 95
       Determine Technical Solutions and Physical Design 96
  What's Next? 100

Chapter 6 — Implementing the Solution 101

  Final Review, Going to Bid 102
  Project Lifecycle 104
       The Implementation Plan. 107
  Testing Procedures — Trying to Get It Right the First Time 110
  Maintaining the Health of Your Solution 113
  What's Next 115

Chapter 7 — Presenting the Telephone Applications 117

  Telephone Services Primer 117
       Inbound and Outbound 118
       Dedicated Services 118
       Centrex 120
  ISDN benefits 122
  The Applications 124
       ISDN to the Home. 124

TABLE OF CONTENTS

    Small Office 129
        Wire Closet 130
        Office Setup 131
        Home Office 134
    PBX-Based Offices 136
        PRI and the PBX 138
        Drop and Insert 142
        BRI to the PBX 142
    Final Example 144
What's Next 145

Chapter 8 — Presenting the Data Applications 147

    Datacommunications Services Primer 148
        Some History 148
        ISDN and Datacommunications 149
        Connection Types 151
        Connection Components 152
            Internal Adapters 152
            External Serial Port Adapters 154
            Network Adapters 155
    The Applications 156
        The Store 156
        Remote System Access. 159
        LAN to LAN Services. 163
What's Next 164

Chapter 9 — Presenting the Video Applications 167

    Videoconferencing Primer 168
        Standards 168
        The Basic Components 169
    The Applications 172
        Conference Room Systems 172
        Desktop to Desktop Solutions 175
        Video telephones 179
What's Next 180

Chapter 10 — Putting It All Together 181

    The Final Mix 182
    What's Next 182

Appendix A — Hardware & Software Resources 185

    Telephones 185
    Data Networks. 186
    Videoconferencing 188

Appendix B — Carriers 189

    The U.S. Local Operating Companies 190
        How to Reach Them 191
        Sample Tariffs...193
            Ameritech..193
            Bell Atlantic 194
            BellSouth 195
            Cincinnati Bell 197
            GTE...198
            Nevada Bell..201
            NYNEX 201
            Pacific Bell...202
            SBC 203
            SNET 204
            US West 205
    The Long Distance Companies 205
    Going international 206
        Tariff Examples 209
            Telstra 209
            France Telecom...210

Appendix C — Other Resources 213

    On-line Resources and Users Groups 214
        Users Groups..214
        General Resources 215

TABLE OF CONTENTS

    Resellers, Software, Consultants 217
    Broadcast Audio 218
    On-line Services Providers 219
Books 219
    ISDN Technology 219
    Network & Systems Design 220
    People and Project Skills 220

Appendix D — The CD-ROM 223

CD-ROM Organization 224
Installing the software 225
    Adobe Acrobat 225
        Macintosh Users 225
        Windows Users 225
        DOS users 226
        UNIX Users 226
    netViz 226
Using Acrobat 227

Index 229

Foreword

The '90s have been consistently exciting times for us technophiles. The so-called information superhighway is coming through, foreshadowing a future of abundant bandwidth and quick access to worldwide data resources. It's a truly captivating vision.

Yes, it will take years to fully flower. But there are things you can do today — practical, economical things — to avail yourself of substantially more information power than you've known to date. It's called ISDN, for Integrated Services Digital Network.

I grant you, it's an acronym.

But in use, it glides like the wind.

Some readers of this book belong to the demographics I call the "Nearly Wired," the 40 million who've read about the Internet and want to go surfing. They've heard ISDN offers about 10 times the surfing power. Naturally, they want in on it. Naturally, they want it "plug and play" easy. And if it's *really* just the Internet you want, the industry is getting close to that level of convenience.

But I believe most people who read a book like this come from the "Hard Wired" demographic. They know an ISDN line offers a lot more power and a lot more options than Web browsing requires.

These advanced users may want to connect multiple devices. They may want to send faxes, download files, and talk on the telephone without having to continually switch things around. Whatever the case, they want to use that extra power of ISDN — beyond Web browsing. They know there are all kinds of configurations and combinations possible and they want to optimize it for their own personal styles of networking.

Why such a book? Because more powerful technologies are rarely as simple as their predecessors, and ISDN is no exception. It is not as easy to order or install as plain old telephone service. We're making progress in that direction, but ISDN is inherently more complicated than analog circuits. It takes a certain amount of effort to learn its possibilities, a certain amount of reading and talking with others before you're ready to harness its full power. That's obviously just what you're doing.

It seems to me likely that anyone reading a book on ISDN probably knows there's a class of information out there that's neither inborn or intuitive. It's often difficult, and you have to dig it out. But you take on that challenge because the information is powerful and valuable. It opens doors to both practical advantage and a better quality of life. And it's fun as well.

This book is full of such information.

Dave Dorman
President and CEO
Pacific Bell

Preface

In the face of massive advertising about using the Integrated Services Digital Network (ISDN) to get onto the information highway, to go home to work, or to be able to run your fax, telephone, and computers all from this one technology, it seems a very enticing and desirable technology. But what happens when you call the phone company that convinced you to do this — they can't do any more than sell you the line, leaving you to find all the hardware and software to make your application work. Where do they get off selling you the application but not the solution?

ISDN, in fact, is a promising and powerful technology. It is mature in the sense of being over 10 years old, yet it is young because the market never took to the technology, and quite honestly, the providers really didn't have the technology infrastructure to make the promise a reality. But now that ISDN is readily available to most people in the major metropolitan areas, with several telephone companies claiming 100 percent coverage, it is a technology to be considered and reckoned with.

What can this technology really do for you? With Basic Rate Interface (BRI), you can get three phone lines down one wire pair, with two high-speed voice/data lines and one low-speed data only line. The Primary Rate Interface (PRI) service provides 23 or 30 high-speed voice/data lines plus a high-speed data only line. So what, high speed or low speed, will that really make a difference to your life? Yes, absolutely!

The speed available for each high speed line is 64 Kbps, over four times the typical 14.4 Kbps modem speed installed today or double the 28.8 Kbps modems. Just that is worth attention and praise because of the highly graphical interfaces in Windows, Macintosh, and UNIX. Add to that the ability to use multiple lines and you can see the speed advantage. But, let's not forget that although the world doesn't evolve around computers, our telephone services can be greatly enhanced with this technology. Since ISDN is digital, our calls are clearer, our faxes are more reliable, and the phone company can provide more flexible and creative services using this technology.

What all this really means is that you can provide more sophisticated services to your customers, to your staff, and interact more effectively with anyone you need to reach by a phone line. With speed, clarity, and flexibility the key characteristics of ISDN in a dial-up world, it is the on-demand technology to move you ahead of your competition, or get you home in time to enjoy your loved ones. So take some time to understand the technology and then put to use the technological solutions described here. Use this book as your guide to defining, selecting, and acquiring ISDN to meet your needs.

■ AUDIENCE

Why would you want to read this book? Take a few minutes and see if any of the following descriptions fit you and your needs.

PREFACE

You are someone who has decided that the promise of ISDN is worth understanding well enough to make decisions about implementing solutions using ISDN. Your interest in this technology will put you in the forefront of investigating and using ISDN, but not on the bleeding edge — it's a technology that is fairly stable and ready for primetime.

As a manager who has people crying out for more time at home, you want to understand the technology issues of distributing people outside your facilities and still providing them a reasonable access to the information resources you've developed. Your technical staff, some advertising, or an article has got your eye and you decided that you have to understand enough of the technology and applications to fit them into your business plans.

Working in the MIS organization, you are someone who strives to enhance their knowledge about technology and how to use it to serve the business objectives. You have users that need you to guide them into better use of their systems, and you want to take on a leadership role in the growing areas of telecommuting, videoconferencing, and possibly some telephony services.

The telecommunications department is getting stretched thin and all of this talk about high capacity lines is driving you crazy. Your boss has come in with a mandate to implement ISDN because it can save you money on your telecommunications costs, but you don't know where to start. On top of that, someone has told you that your data guys over in MIS have been throwing some of these lines in and you haven't understood why.

As an end user, you have been wanting some practical way to get home a few days a week to get your work done, but you're stuck because the best the company can do is give you a slow modem that doesn't support all those wonderful programs that MIS came up with. You heard that ISDN was the high speed way to go, but you can't seem to raise the awareness in MIS or the telecommunications department — time to get the answers yourself and explain to them what you want.

You have just started to work for a high-tech company selling products that support distributed applications, letting small offices talk with their corporate big brothers. You know that many of these prospects will need to understand the technology to interconnect them in an on-demand environment, and ISDN seems to be the ticket. It's time to find out enough information to explain the subject, and perhaps have the next few applications in mind for that long term account growth.

You can see from this small sampling of situations that many of you will benefit from this book. This book is intended to stimulate your knowledge and ideas on using ISDN to better your business and your life; giving you enough knowledge to tackle the small projects on your own, and to know enough to bring in the big guns for the larger and more complex situations.

■ ORGANIZATION

In the beginning, as always, you start with a basic overview of ISDN, what you can do with it, and where it is headed as a technology. Chapter 2 will provide the in-depth information on the structure and functionality of ISDN, giving you enough information to be conversant with the really technical folks who make the whole thing work, without making you a technical expert. In Chapter 3, the different types of hardware you need to connect computers, phone systems, and videoconferencing equipment to an ISDN line will be explained. Next, it is time to understand how the telephone companies, or carriers, speak, talk, and walk the world of ISDN. Chapter 4 will guide you through their world and help you take on that challenge.

You're almost ready for the applications, but still need a few tools under your belt to take advantage of them. Chapter 5 will give you a short lesson in design applications that use ISDN, taking you from the initial business problem definition through the construction of a document to bid the project, and then onto the process of getting what you want from the vendors. With product in hand, it's time to implement your project, and Chapter 6 is there

PREFACE

to give you the foundations of an implementation plan to cover your management of this process.

Now you are armed with the understanding to work with the applications presented in the next four chapters. You start with telephones in Chapter 7, looking at a variety of telephone system interconnection; and enhanced services provided as a result of the capabilities of ISDN. On to the computers in Chapter 8, where the many examples will take you from interconnecting your offices to working at home. Being able to see people at the other end of the phone line has been an intriguing promise since the early 1960s and the first prototype video phones. Chapter 9 will guide you through the many examples of how ISDN makes this possible through videoconferencing. A final note is offered in Chapter 10 to pull everything together.

Your guide to resources will come in the appendices. There you will find lists of suppliers, products, telephone carriers, and other books or resources to help you refine and enhance your knowledge of ISDN and its application. The final appendix covers the CD-ROM contents and how to get started with that resource.

■ CONVENTIONS

You have several key visual aids in the book to help you zero in on special information. Occasionally, you will find a definition element, shown below, again to the left of the paragraph where that term is used. This element should keep you from pulling a dictionary or bouncing through the book to understand a new term. One other element will be the shaded boxes where various examples or pieces of information are separated from the main text. These boxes may not have labels, but will always follow the paragraph they have been mentioned in.

Definition Box   **Example Box**

There are icons that will appear to the left of the paragraphs that identify special information that is important to you. The CD icon will tell you that additional information on this topic is on the CD-ROM. Every once in a while, you will find four other icons that will prove invaluable to you. The first is the Caution, where I will give you a helpful piece of information to steer you clear of problems in the process. The second will be the Bright Idea, where I will stimulate your ideas with brainstorms of my own. The last two icons are for applications: the letter is information from a User Application Story, the person is a Sample Application.

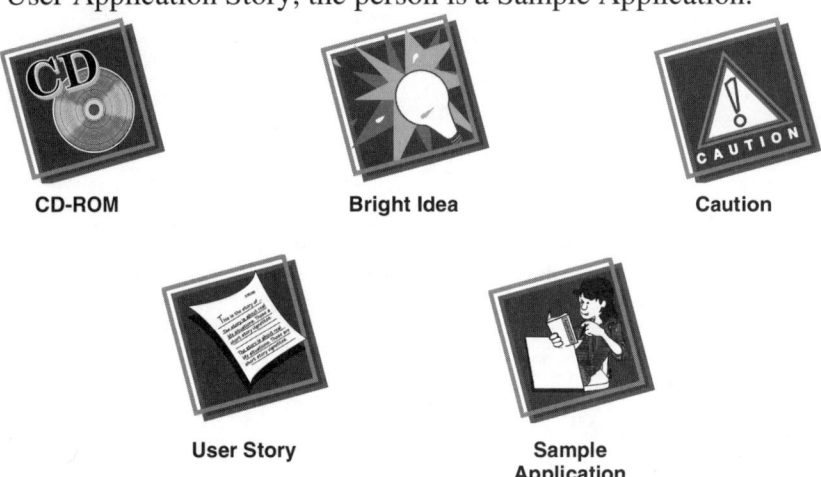

CD-ROM   Bright Idea   Caution

User Story   Sample Application

The CD-ROM included with your book has several exciting features. The first is the inclusion of the complete *Catalog of National ISDN Solutions for Selected NIUF Applications, Second Edition,* made available courtesy of the North American ISDN Users' Forum. This catalog gives you hundreds of additional pages of detailed technical information to take you further than the material in this book. Just as exciting, this book is reproduced on the CD-ROM, with some additional materials for your use in your ISDN project. To make these materials useful in electronic form, the two books and forms have been formatted and stored in Adobe's Acrobat format, and the reader software is included on the CD-ROM. Appendix D will provide you the instructions to install and get started with the Adobe software.

PREFACE

■ ACKNOWLEDGMENTS

No project is ever accomplished without help, and I have received tremendous support from a number of people who have my thanks and gratitude for making this a successful and pleasurable project. I have to start with Karen Gettman, my editor, who was willing to review my proposal and see the value and need for the book. Pat Pekary, at Hewlett-Packard, for wanting to have HP sponsor the book, and Ken Myers for championing the book at HP. My personal experiences with ISDN are shaped by my local phone company, Pacific Bell, and Tom David was instrumental in getting me the support I needed to *play* with many of the different configurations of ISDN, to make the applications work.

Additional thanks go to: Art Slaughter at Pacific Bell for technical assistance and education; Marlene Larson at AT&T for getting me additional information on the international aspects of ISDN; and Dick Slezak at AT&T for insights into international ISDN and some very innovative applications. Thanks also to the folks at Prentice-Hall: Barbara Alfieri, Sophie Papanikolaou, Gail Cocker, Camille Trentacoste, and Mary Sudul, for making the construction of the book much easier than I would have; to Barbara Danzinger for the excellent copyedit work; and to Talar Agasyan and Jerry Votta for the creative and very appropriate cover design. Bob Lanphar at Votaw Data Systems; Mike O'Gorman of Mike O'Gorman & Associates; and Margery Mayer at TeleConsultants Inc., for reviewing the book and providing me with valuable suggestions and improvements.

The following list of companies graciously loaned me equipment, time, and expertise to bring you accurate information and innovative ideas. Thank you to: 3Com, ACC, Ascend, AT&T, Bell Atlantic, Bellcore, BellSouth, Cisco, CoreLogic, Digi, GTE, Hewlett-Packard, IBM, Lucent Technologies, Motorola, Pacific Bell, US Robotics, and Vivo. Special thanks goes to PSInet for working with me on their ISDN access to test out many of these products.

CHAPTER 1

Why All the Hoopla?

It is exciting to understand a technology and get all jazzed by what they tell you it can do. But you won't see any real benefit until you translate the technology into applications that you can implement. What are the real life challenges you face and how can this technology help you? This is the question you must answer, and to help you get there, this book will explore the technology and the many applications already implemented or envisioned because of ISDN.

■ APPLICATIONS — THAT'S WHY!

How can a technology, one that simply digitizes your telephone service, enable so many new ways of doing business and communicating? It results from the fact that today's information can be reduced to digital form. However, the data are so voluminous, it requires great speeds to transmit it in a timely fashion.

ISDN offers the first affordable, dial-up, technology to support this requirement.

Just look at what has been turned into those bits (remember, a bit is a single one or zero) of data. Our voices have been transmitted by digital encoding since the 1960s, while computers have been doing business in this way for over 50 years. Even the concept of Morse code resembles digital transmissions, where each letter is coded with a sequence of dots (ones) and dashes (zeros), and then sent over a wire as tones of different durations. Now we can encode our video transmissions into bits, but even the new techniques to compress the long strings of bits necessary to encode a single picture (a single screen shot on a computer monitor is 304,200 bits, without color information) have left a great demand for speed.

Digital voice is neat, but what difference does it make to your business needs? By itself, it doesn't, but the foundation of how your voice travels as a digital message does. ISDN is built on information controlling the call traveling separately from the data for the call. The call controls travel in the form of messages, and through this message capability you are able to add all sorts of new services. Through messages, you can program your phone, or any other ISDN device, to perform any function that you want, as long as you meet the signaling requirements within the ISDN standards. With video images being converted to bits, and computer data already in that form, ISDN can transmit everything without needing to be aware of the content.

■ WHAT IS ISDN?

Integrated Services Digital Network, it's not easy to remember; maybe that's why it took so long for it to catch on. But for most of us in the digital revolution, ISDN is touted as the high speed on-ramp to the information superhighway. Is it really all that it is marketed to be? No, not really (does anything really match the marketing promise?). It does, however, solve real problems with flexibility and speed that you can't get at the same price and performance levels in any other technology today, and that's what makes it so exciting!

Simply put, ISDN is the ability to pass multiple phone lines down a single wire, and by doing so in a digital format, open the door to large data applications in the domain of voice phone calls. But how did this come about, and why does it help to understand the history of ISDN? I'm going to give you a brief history on how the standards were set and determined the ultimate future for this technology. Those decisions, made over 35 years ago, have given ISDN a finite life span, and helps you to understand what capabilities and challenges that ISDN has in the next decade.

Evolution from the Telephone

Telephone services began in a world of point to point connections that were very manageable, where operators could use the cordboard panels to directly connect a call from one person to another. This action completed the single-wire pair needed to connect two telephones. But, this method had severe limitations. In the first place, you had to increase the number of people to handle calls as the number of subscribers went up, something every business abhors. Second, when a caller requested a call be placed to a type of business, the calls were often routed to the business choice of the operator, not the caller. This latter situation actually led to the development of the first automated phone switches, by a company that wasn't the preferred choice of the local operators.

Going Digital

Things were working fairly well until the 1950s, when the technology was evaluated for its ability to handle future capacity requirements. The telephone companies recognized the need to increase their office to office facilities to handle even greater numbers of calls. Somehow they had to make the wires handle more than one call at a time.

Going digital was the answer. Time was spent understanding what it would take to transport a reasonable representation of the human voice without completely duplicating it. In an analog world, every component of a signal is maintained throughout a system, while in a digital world it is composed of ones and zeros(bits). In Figure 1.1 you can see the wave form of a recorded voice in the Sound Recorder picture, while below it is a representation of the signal encoded into the stream of ones and zeros (a stream that takes thousands of bits to encoded the voice sample shown in the Sound Recorder picture).

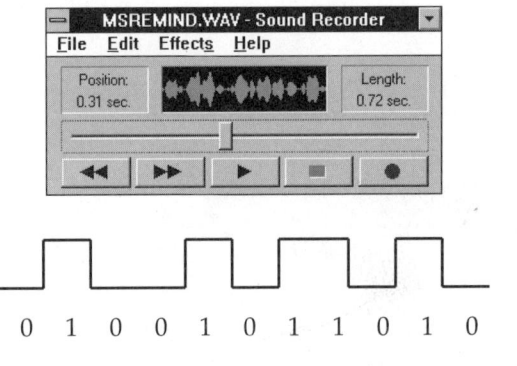

FIGURE 1.1 ANALOG AND DIGITAL SOUND FORMATS.

While the human voice has a normal frequency range much larger than 4 Kilohertz (KHz), it was found that this limited range can be used to reproduce most sounds that a person makes. In order to reproduce this range, engineers identified that the sampling rate (the number of times a signal is measured) for the human voice would need to be twice this range, or 8,000 times a second. It was also decided that eight bits of information, representing the sound level and frequency, would be recorded for each sample. This decision gives us the 64 Kbps speeds for each channel of ISDN we hear about today. Then the engineers determined that you could divide a telephone signal into time segments, and into these time segments you could insert the

> 8,000 samples/sec
> 8 bits/sample
> 64,000 bits/sec

Eight bits make up a byte.

Kbps is Kilobits per second,

sound bytes. If you increased the number of time segments on the line to a multiple of this 64 Kbps rate, you could interleave the sound bytes and simulate multiple phone lines.

Now the telephone network was digital between their main transmission centers but still operated on an analog basis from the central office and your equipment. By the 1970s, the technology had evolved far enough that it was decided that the digital transmission methods used to transport the sounds from office to office should be moved closer to the customer's premises. Moving the conversion to digital form to the customer's premise improved the transmission quality on that part of the circuit and allows the telephone companies to build more effective networks.

> **Premises —**
> **your location.**

With digital facilities available to the customer's location, two primary classes of applications emerged. The first was the transmission of computer data in digital formats, at speeds of 56 Kbps to 1.544 Mbps. On the voice side of the applications, the T-1 circuit, 1.544 Mbps, provided 24 channels of voice circuits. While voice and data could be carried on any channel, the function of a channel was static, assigned when the circuit was installed, and required equipment and telephone office changes to switch between voice or data functions.

> **Mbps —**
> **Megabits per**
> **second**

Enter ISDN

The primary technology standard to enable switched digital transmission of voice and data calls across the same wire pair is ISDN. In the late 1970s announcements were made that this technology would bring several flavors of speed to access the network: Basic Rate Interface (BRI) at 144 Kbps and Primary Rate Interface (PRI) at 1.536 Mbps on a T-1 or 1.984 Mbps on a E-1 circuit. Outside the U.S., Canada, and Japan, the E-1 is analogous to the T-1, with a total line speed of 2.048 Mbps.

Great excitement has moved through the industry as people recognized the possible applications that this technology can support: videoconferencing, high-speed computer access, more phone calls down a single wire, and shared voice and data use of the circuit. So what happened, that was over ten years ago — why doesn't everyone have this technology now?

■ VARIATION ON A THEME — THE STANDARDS

Standards drive the availability of choices in a marketplace. Through standards, the telephone industry has enabled you to buy equipment from hundreds of manufacturers, knowing that if the equipment meets certain standards, it has a high probability of doing what you need it to do. Think about just the telephone itself. You know how to operate the phone because all phones have some kind of dialing mechanism, with the same basic keypad layout of ten digits and a pound and star key. Picking up the phone initiates a dialtone, which is recognizable worldwide, and the busy signals or rings you get tell you if the call is going to connect or not. All this happens because of standards.

On the dark side of standards comes the holes in the specifications, the ambiguity of the language (sometimes put there on purpose), the delays that a standards process introduces, and the jockeying for a particular approach to a standard that will favor the party that developed it. But in the end, it is the standard that opens up the market and lets you benefit from the competition it supports. It's useful to recognize that standards really begin as a proprietary technology or design that an individual company then promotes as the logical choice for everyone to follow as a standard.

ISDN has standards that govern its implementation in countries worldwide. Ultimately, all standards are passed up to the International Telecommunications Union (ITU) where member countries define the international standards to ensure communications on a global basis. Here in the United States, it is the Telecommunications Industry Association (TIA) and the Alliance for Telecommunications Industry Solutions (ATIS) that guide the standards. In Europe it is the European Telecommunication Standards Institute (ETSI) that guides the development of standards across countries and feeds recommendations to the ITU.

It was in November 1992 that the first calls were made using National ISDN-1. This marked the first demonstration that ISDN had passed the proprietary phase of the technology and that all of the carriers and equipment manufacturers were ready to make the technology accessible. Work moved ahead on National ISDN-2 and -3. It was hoped that by now National ISDN-2 would be the prevalent standard, but implementation is behind schedule and is delaying the implementation of National ISDN-3. The primary delay in bringing the standards to market has been the cost and scale of upgrading the many central office switches that will provide this service. The standards will be covered in more detail in Chapter 2.

> **National ISDN is a United States and Canadian series of ISDN standards.**

■ APPLICATIONS — THE PROCESS

It's the use of ISDN, the actual application, where the beauty and promise of ISDN is realized. By thinking through how you will benefit from the distribution of people and resources, bypassing the delays of travel and conventional means of communication, you will develop your application using ISDN.

Rising Expectations

It is our ever escalating expectations of technology performance that drives us to ISDN and other future technologies. Think about the past, where it took days to weeks, or even months, for messages to travel back home. Then along came the telegraph, and with it digital information, in the form of Morse Code, which was able to travel great distances in minutes or hours. Next came the age of telephones and radio, where our voices traveled at the speed of light, and we could now move information in seconds across the globe. Now we are used to seeing conflict, disaster, and celebrations, all within minutes of their occurrence, from anywhere on the planet.

On a simpler scale, it used to take 10 to 30 seconds to have a dialed call connect to the other end, and for you to get a ringing signal. With touch tone phones, that time is down to a few seconds. With ISDN that time drops to as little as 300 milliseconds. Even so, our appetite for speed escalates, as our comfort level and expectation rises with each new level of performance.

Designing Applications

Designing applications is the heart of this book and understanding your requirements should be the heart of your work before you pick the technology to solve your problems. There are many questions to be answered as you decide whether ISDN or some other transport technology will meet your needs. As you look at your business problem, think about the flow of information, and in this context, I mean data, voice, and video. Consider how much information needs to be moved, how quickly it needs to be moved, and where it is needed. Each of these primary questions will guide you to consider or drop a given technology.

The answers to these and other questions will become the framework of your model and basic requirements definition. By defining the characteristics of your needs, you can match each technology and capability to see what will work for you. This will give you a better insight into what technology to try (conduct a trial) for your solutions. This process will also help you scale your solutions and trials to a size meaningful enough to prove the technology, but small enough to give you a chance to make a mistake and still recover.

In Chapters 7 to 10 you will find many examples of applications with the details to help you understand and plan for implementing your own solutions. Following are some samples of what's ahead.

> **Trials are a period of testing of a technology or solution without making a commitment to implement anything after the trial period.**

> ### Retail Shop
> How about the simple task of operating a small retail shop? You sell craft items, using a simple cash drawer to do business, and never had a lot of business; that is, until the community was selected as a site for an Olympic event. Now you know there will be thousands of visitors coming every day and you are planning to sell all of the souvenirs you can. The process of calling in all of your credit card authorizations is fine for now, but won't be when you have a dozen people waiting in line, what will you do?
>
> ### Branch Office
> Your branch office, out in that resort area where everyone wants to live, runs on an old phone system you bought back in 1980. Your MIS folks want to change out that leased line running at 9.6 Kbps; you want to add voice mail and disaster recovery to the phone services. Good fortune has smiled on you because the phone company just upgraded their switch and announced that ISDN is available. What's your next step?
>
> ### Telecommuting
> Los Angeles, the place where everyone has a car, and no one lives near their office. With wide open spaces, millions have fled to the suburbs for more affordable housing, but those suburbs are 50 miles away and hours of travel by car. Your employees are among those millions with hours of daily commuting time. Enter ISDN, the advertising says that every department can build a telecommuting plan to cover phone services, data, and videoconferencing. Are you going to take the bait and try out this technology?

■ STATE OF ISDN

How available and ready is ISDN? More so than you might think, but a lot less than you might need. The technology has been the talk of the town for several years, with every carrier putting on their best face to tell you that the technology is here, it's happening, and they can give it to you right away.

Reality has since set in and you will have to deal with the marketing hype versus the reality. The reality is that many of the carriers can install ISDN, but only in limited areas, or in a longer than expected period of time. With that disheartening news out of the way, let's look at what's really happening out there.

At present many of the Bell operating companies report 100 percent availability of ISDN in their territories. What this really means is that where the service is not available, the carrier will backhaul (run a line from your central office to another) the line to a central office that has the proper equipment. In the past, the cost of this service was increased by the distance of the backhaul. Many tariffs bury this cost, allowing the telephone companies to seed the service in new areas, until critical mass is reached to upgrade the local facilities. In Figure 1.2 you will find the general territories of each of the carrier companies. With the increasing possibility of mergers, you can expect this map to change dramatically. As this book went to press, SBC and Pacific Telesis announced a merger, which combines SBC, Pacific Bell, and Nevada Bell. Other mergers may not be far behind.

National ISDN-1 or the original custom ISDN services are rolling out in many areas, but are not 100 percent available yet. Prior to the implementation of National ISDN-1, the implementations of ISDN were custom or proprietary. Each equipment vendor took a different approach, and interoperability between switch vendors was questionable, just not possible. With National ISDN-1, the interoperability issue went away. But AT&T had implemented features in their ISDN model that exceed the national standards, so they choose to continue a custom implementation that exceeds the standards whenever possible.

National ISDN-2 is beginning to roll out, and National ISDN 3 will begin to appear when each telephone company is ready with the software, hardware, network, and marketing plans to support the services available under this standard. Many of the issues of implementing these standards relate to the support and billing infrastructures, as much as they do to the software and hardware to provide the services.

> **Central Office** — the first point that your phone service enters the phone company network.

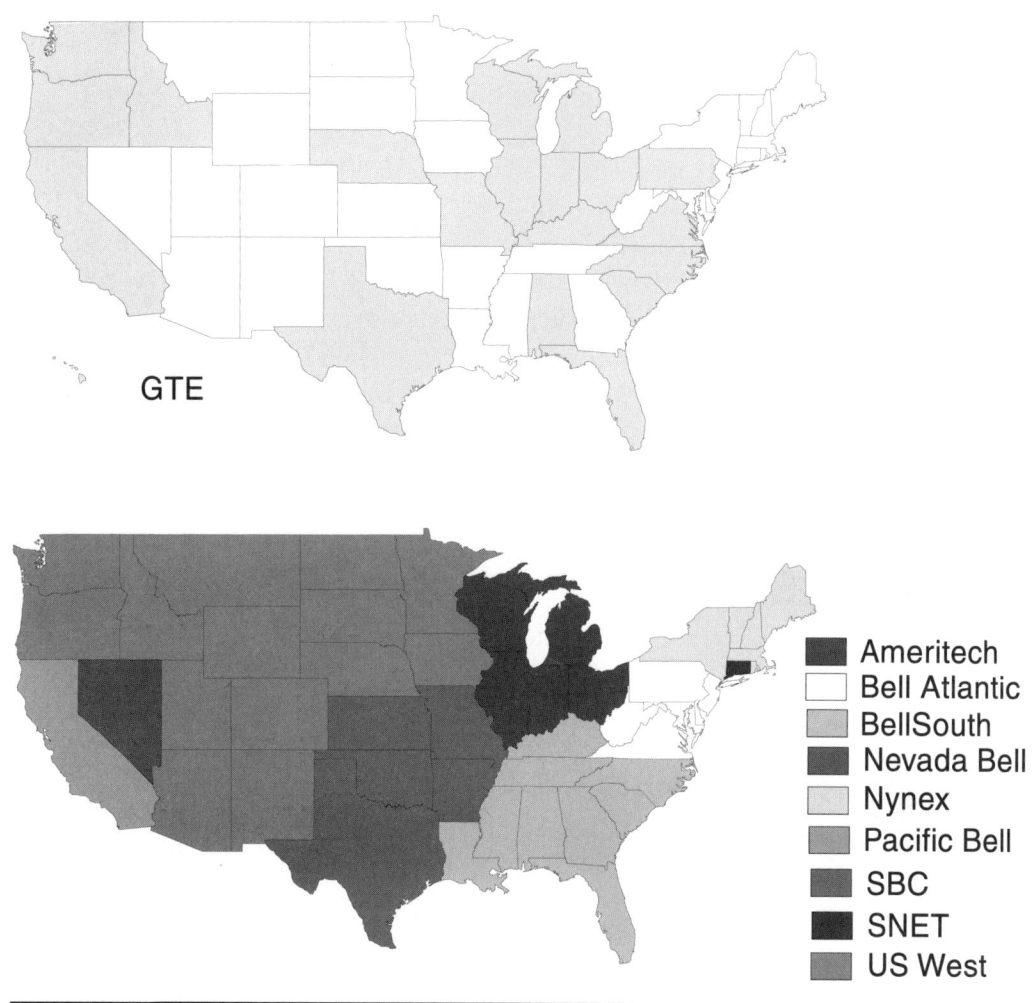

FIGURE 1.2 REGIONAL TELEPHONE COMPANY TERRITORIES.

In Europe the ISDN technology is being rolled out under the Euro-ISDN banner, which is driven by ETSI. These standards are being adopted across country borders to improve the interoperability between countries, much as National ISDN is solving the problems between equipment vendors and telecommunications carriers in the United States. Within each country the implementation of ISDN is decided by the regulatory agencies there, making it necessary for you to understand that every service or equipment capable of ISDN may not be available because of decisions within these organizations.

But what good is a line without equipment to connect to it? One of the most exciting aspects of the technology is the drop in equipment prices this past year, and the predicted reductions for the coming years. Only a few years ago you would find ISDN adapters nonexistent for computers, ISDN phones were $700 or more, the interfaces to your phone systems might run $5,000 or more, and the lines were thousands of dollars. Now you can find adapters for computers at less than $250, phones in the $300 range, and manufacturers discussing chips to put on computer motherboards for $35 each. This kind of pricing is bringing ISDN equipment into the price threshold that most people can afford.

With most tariffs dropping into the $30 to 50 range for the home market, up to several hundred dollars for business BRI lines, and PRI lines going for as low as $600, ISDN is finally hitting that magic break point where a market will take off with significant demand. And the more people that jump in, the cheaper the service is going to become.

■ EMERGING COMPETITORS

There is always something down the road, and these days, along it also. Two very hot topics in the industry are Cable modems and Digital Subscriber Links (DSL). In Figure 1.3 you can see the enticement these technologies represent. Both have some significant advantages over ISDN for data applications, but have challenges to assume the role that ISDN can provide today. Let's take a look and see what promise there is, and what challenges they have left to reach their promises.

Cable Modems

Cable TV has penetrated many of our homes and in 1995 it was estimated to have reached 59 million households in America, and goes by 82 percent of all homes. The high speeds available on

WHY ALL THE HOOPLA?

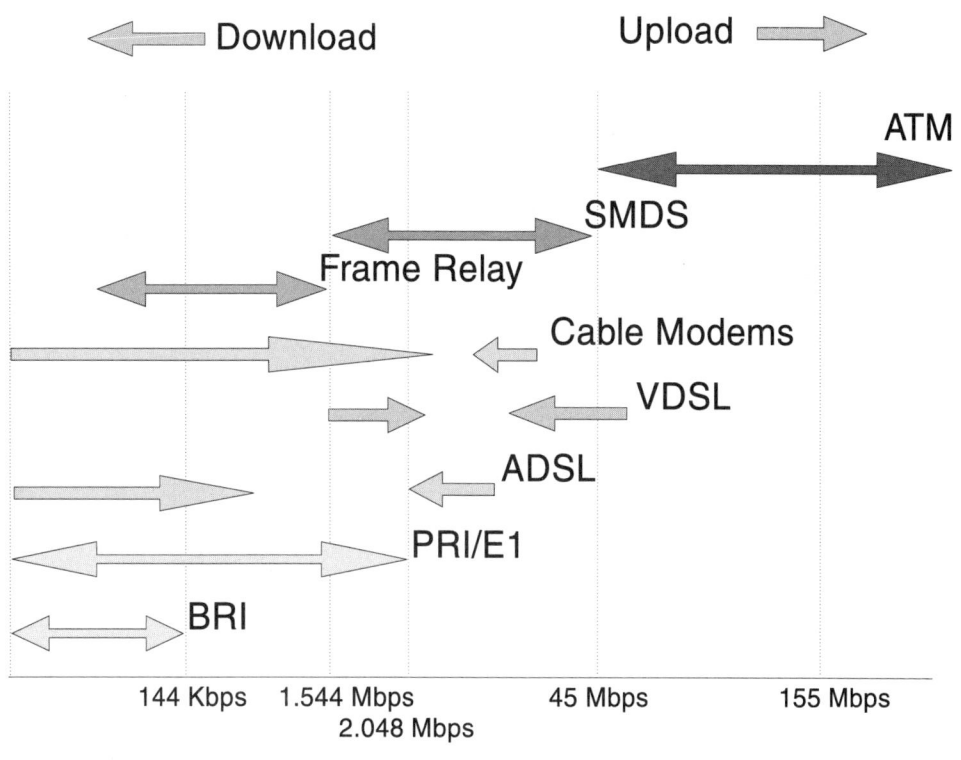

FIGURE 1.3 BEYOND ISDN, HIGHER SPEED TECHNOLOGIES.

this cable, anywhere from hundreds of kilobits up to 40 Mbps, makes this a very attractive technology to deliver large amounts of data to the home. These modems have been designed to deliver the video content in digital form, and by being digital, they can carry other digital data. In a market where users are downloading large amounts of data, pictures and video, this design model works well; because the request (the upload need) for information or video takes very little data or speed, while the download requires more speed than ISDN can deliver today.

In order to transmit this large volume of signal across a long distance, devices called repeaters are installed on the cable (fiber or coax) to strengthen the signal at various points in the network. The repeaters installed in 95 percent of the cable systems today are designed to carry a large signal from the head end, or origination point, and very little in the other direction. This model precludes large file transfers, videoconference, or other large, speed intensive applications that originate at your site from using the cable modem system for transport. Figure 1.4 shows the basic structure of a cable system and highlights the role and large number of repeaters that the cable network requires.

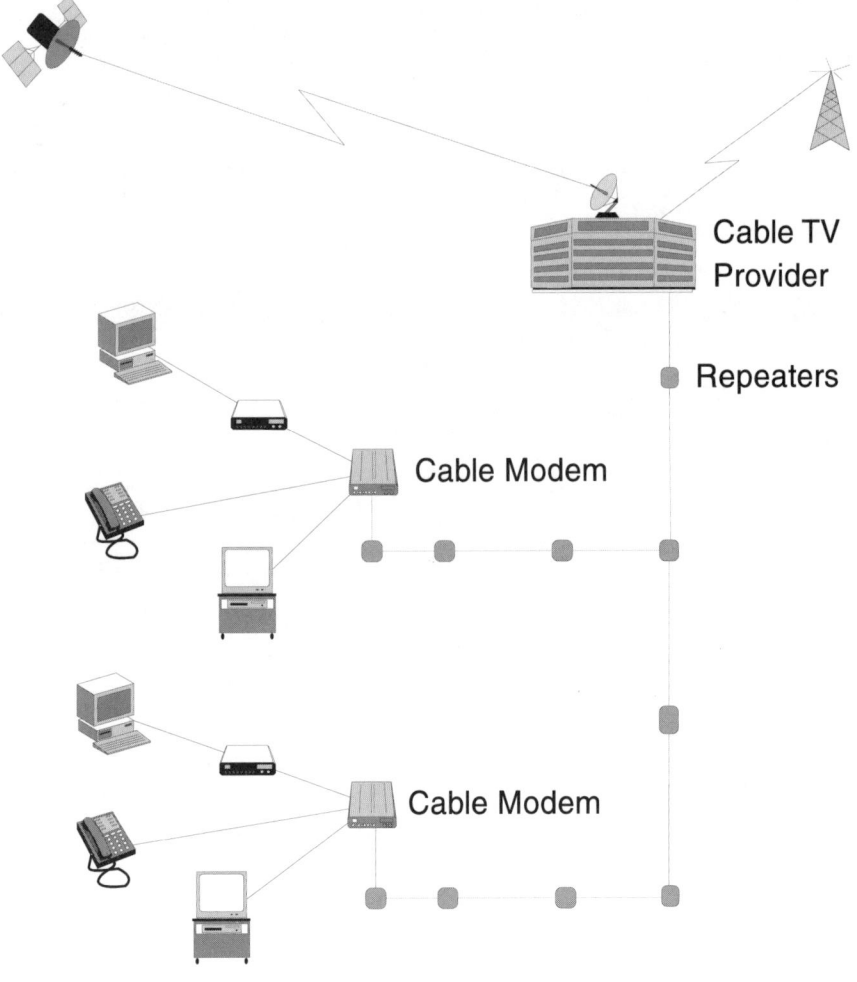

FIGURE 1.4 CABLE TV SYSTEM CABLING INFRASTRUCTURE.

WHY ALL THE HOOPLA?

If the infrastructure will support large amounts of bandwidth in one direction, why is bidirectional support so important? Many of the exciting opportunities on the Internet and in videoconferencing are bidirectional applications. The next example highlights the need to provide data to the Internet.

> You have a hand-dyed fabric company where you have been selling your fabric at quilting shows, and you've stirred up a lot of interest in your products. Lately, you have been *surfing* the Internet and found several quilting newsgroups and other resources where some of your competition is hanging out and gaining business. Your fabric designs are best shown in color, and you have other pictures and information you want to share to increase interest in your products and services.
>
> Being in one of the early adoption areas, your Internet connection is coming down your cable TV line, but right now you know that all requests to the Internet are going out your phone line at 14.4 Kbps. But you need at least a 128 Kbps circuit, with committed bandwidth, to give a reasonable response time to requests from the people visiting your Web site. Now you have to get a separate line for the server instead of using the same circuit for all of your Internet access. The cable TV model no longer fits your immediate needs.

Bandwidth — a common expression of the speed or capacity of a circuit.

To provide for these capabilities in the short term, the cable TV companies will have to install their own telephone systems and data networks, and in the longer term, increase the number of link devices to support substantial upload activities.

The cable TV providers are building their systems and technology around shared bandwidth model. The model assumes that of every 2,000 homes that might be served by a cable modem subsystem, the system will provide for 30 to 40 Mbps of download and 12 Mbps of upload speed. The next part of the assumption says that if things go well, there will be a 10 percent penetration of this group, so only 200 homes will be sharing the bandwidth, and that a fraction of those will do so simultaneously. If you assume that just 10 percent of this last group are active, then you have to divide the bandwidth by 20 to get what amount of bandwidth each home has available for its needs, which on average is 1.5 Mbps download, and 150 Kbps upload. These are some big assumptions to give you an upload speed barely above BRI and a download speed equal to PRI.

The cable TV industry acknowledges that their platform is built around a content delivery model, not a bidirectional use of the system. The telecommuter, videoconferencing, and telephone models don't fit the current generation of cable modems, there is too much bidirectional activity. Each of these applications will require other equipment and technologies to use the same cable infrastructure.

The amount of bandwidth a coax cable is significant. Because so many homes have access to this cable, it brings to many people a choice in who will supply them their communication services. But, for now, that full service alternative is still years away.

Digital Subscriber Line

The DSL technology has been around for a number of years. The current generation of 56 Kbps and T-1 services use this basic technology. There are three variations of this technology that you should be aware of, because trials are in process and some commitment has been made by the telephone companies to deploy this technology.

The basic DSL technology is High-bit-rate Digital Subscriber Line (HDSL) which provides for 768 Kbps of bidirectional speed, six times faster than ISDN. That's great, but here's the catch. The modem to support this will sell for about a $1,000, and you need one at each end. The telephone company has to install special equipment to support this line, very similar to the T-1 lines today, and those installation costs are up to several thousand dollars, and lastly, the phone company will have to tariff the service, a process that takes time.

The next generation of DSL is Asymmetrical Digital Subscriber Line(ADSL) and Very-high-rate Digital Subscriber Line (VDSL). These are two very exciting technologies. Looking

WHY ALL THE HOOPLA?

> Surfing — that traditionally male habit of switching channels frequently. This term applies well to the way people use the Internet.

back at Figure 1.3 you can see bandwidth going from 1.5 Mbps to as much as 52 Mbps, enough speed to satisfy most of us in even our toughest video channel or web surfing mood. What else makes this exciting is the fact that the service can be delivered down the same phone wire you have in your house or business today. With all this going for it, this must be a slam dunk for winning our hearts.

Again it is necessary to look at the model and to understand two things, does the transport system fit our model of information or communications movement, and then, how much of the infrastructure has to be changed, and how soon will this technology be deployed. These questions pose the biggest stumbling block to any new technology, and ISDN was itself in that position over ten years ago. The technology is not currently deployed by the phone companies or network providers. Trials are beginning here in the United States, and Telia (the Swedish PTT) has made a commitment to the technology. Even when a commitment is made, it will take tremendous effort, issue resolution, capital, and time to fully implement this new technology.

Conclusions

The solutions today for both DSL and cable modems are for data only applications. Mixed telephone and data applications are not available today. Based on the issues of installing and implementing large-scale voice switching capabilities, it is unlikely that the cable companies or any other new company will be in a position to offer wide scale coverage or services in the next year or two, but this is something that will emerge within several years, especially if there are successful mergers of cable and telephone companies.

If it will be a few years before any company is ready to go beyond trials in specific neighborhoods, then what can you do today? ISDN has been the technology the phone companies have committed to for over a decade. With that commitment has come massive upgrades of their infrastructure, planning for new switches, new networks, and new capabilities to deliver the technology to the market. And since this technology is part of the structure that will take us to the ultra-high-speed networks that are being deployed today, ISDN fits nicely into any migration strategy from today's analog environment.

■ WHAT'S NEXT

Now that you have a brief idea of what ISDN is, what it might do, and how it is evolving, it's time to get some in-depth understanding. The next chapter will take you on a technical tour of the technology, not so much to get you oriented to the bits and bytes of what ISDN does, but instead to give you an understanding of the power and capacity. Use this information to help you know how far the technology will take you and what limitations it might have.

CHAPTER 2

Explaining the Technology

During the course of this chapter you will read about a number of details, some technical, in order to gain an understanding of the capabilities of ISDN. It is truly difficult to develop solutions incorporating ISDN if you don't use this understanding to determine what the various service providers and equipment manufacturers can deliver to meet your needs.

ISDN is a technology that can direct what action needs to be taken to move digital information through any one of the individual paths in an ISDN circuit. Because of the intelligence in the ISDN network, and the fact that all of the information is in digital form, this service can provide almost any type of feature we need. Take the time to understand this flexible nature of ISDN, and then use that knowledge to put it to use for you. In the following sections, the BRI and PRI versions of ISDN will be explained. In the last part of this chapter, the broadband ISDN technology will be discussed.

Maybe the place to begin defining ISDN is to use the analogy of a pipe. Just as pipes come in different sizes, so does ISDN. The key difference being that ISDN has only a few sizes to choose from today, and the real pipe, the physical wires the ISDN service is delivered on, are actually a single circuit. ISDN then divides that pipe into a number of logical pipes, each with the capacity of 16 or 64 Kbps transmission speeds.

In Figure 2.1 you can see that ISDN has the capacity to far exceed the fast speeds of today's 28.8 Kbps modems. This is one of the primary reasons why ISDN is exciting for your data applications. But the technology also handles voice calls, more than one per logical pipe, giving you another powerful reason to use the technology.

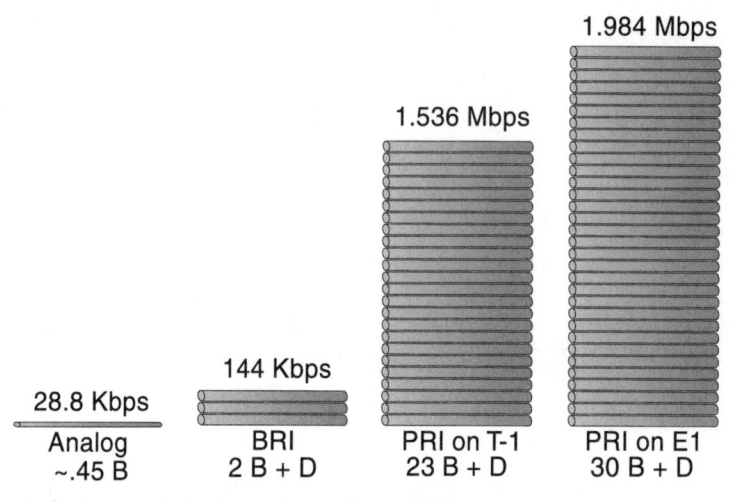

FIGURE 2.1 RELATIVE ISDN AND ANALOG MODEM RATES.

■ NB+D — WHAT'S INSIDE THE ISDN PIPE

nB+D, some kind of strange math? In a sense, yes it is. This formula describes the number (n) of bearer channels (B) plus a single data channel (D), that an ISDN pipe provides. The B chan-

Explaining the Technology

nel carries voice or data packets, while the D channel carries signal messages and data packet information.

Your interest should lie mainly on the number of B channels that a circuit contains. This determines the total speed or simultaneous activities that an ISDN circuit can provide. Each B channel has the capacity of 64 Kbps, and each ISDN circuit has (n) times 64 Kbps capacity. One other key feature of ISDN is the ability to create a high speed, dedicated path between any two ISDN locations, giving the same flexibility as Switched 56 Kbps service, but at a much lower cost and higher speed.

> Switched 56 — an older digital modem technology that supports primary data.

Phone Company Facilities

It's helpful to review the structure of the phone company's facilities, so you understand how a call is carried from your location to another. The phone network I describe is a generalization of the actual networks, which are tremendously more complex than I have room to describe in this book.

> CO is the phone company's central office.
>
> LD means Long Distance.
>
> Tandem is a cross connection switch to link lines from different CO's.

As you can see in Figure 2.2, there are a number of paths a call can take, and many different phone switches that are responsible for making this happen. In a sense, the fact that so many pieces of equipment are involved, it is amazing that we get such highly reliable service from our phone networks.

Path A in Figure 2.2 shows the way a long distance phone call travels, where your initiation of a call will pass through seven or more connections before the telephone network can determine if the call can be completed. If it can, then the network will establish a dedicated path for the call and begin ringing the distant end, at the same time a ringing signal will be passed to you from your local switch, not across the entire network. After your call is complete, this dedicated path is disconnected and made available for another call.

The way the majority of the phone networks communicate is through a dedicated messaging system called Signaling System 7 (SS7). This system allows for the tasks associated with

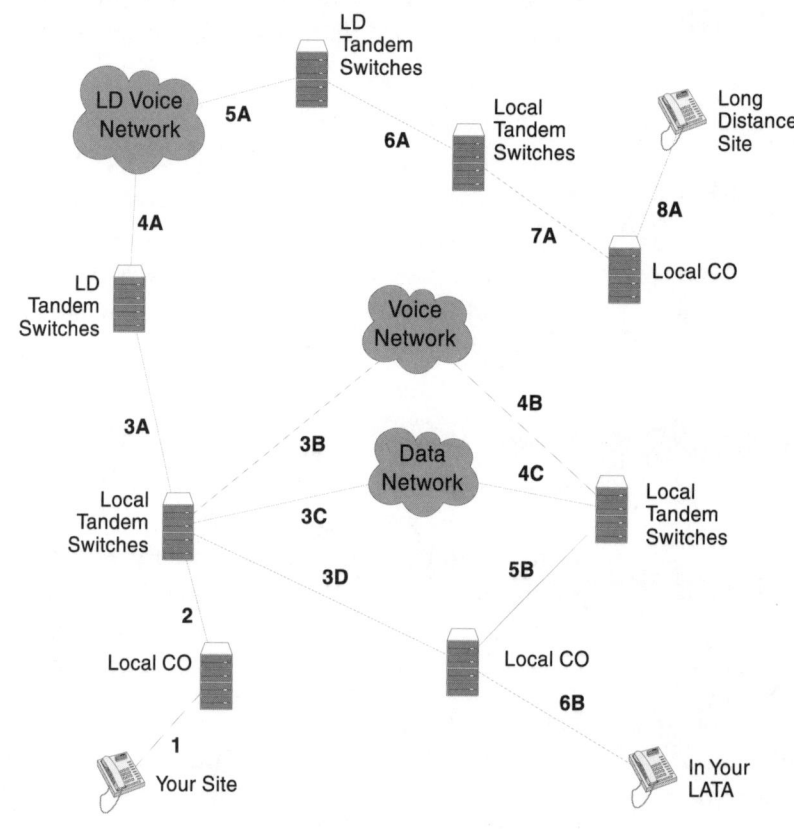

FIGURE 2.2 PHONE COMPANY NETWORK.

establishing, maintaining, and disconnecting a call, to be accomplished without tying up a circuit before the person on the remote end has answered the phone. The SS7 network is separate from the voice network. Not all phone networks have converted to SS7, which means the call control messaging done with SS7 must be done as part of the line the call will use. This use will reduce the total speed available in a circuit by 8 Kbps, to accommodate the messaging tasks.

For a local call (paths b, c, d), the call will travel through the same types of switches and functions, with the variation being based on two things: the call type, and whether there is a direct

Explaining the Technology

connection to the remote office you are placing your call to. The phone company may route your call through a voice network or a data network. The data network is sometimes referred to as an overlay network; because it logically sits on top of the voice network; however, the traffic never mingles.

The switches in your local central office are designated class 5 and handle all of the duties of monitoring the equipment on the lines, and responding to the requests to set up, maintain, and tear down calls, both voice and data. The switches that are identified as Tandems are considered class 4 and handle the complexity of cross-connecting your call to its destination, through the most direct path possible.

One of the important aspects of ISDN technology is its use of a dedicated path for the duration of the call. This impacts the phone company's facilities because that path will remain unavailable for other traffic, even if you don't send any traffic down it for hours. It is the same with your voice calls today, and with the dedicated circuits you purchase. Because of this, the phone companies still have to engineer enough circuits to accommodate this wasted resource. This is a major factor in determining the rates for ISDN.

It is interesting to note that the line card in the central office switch, the phone company's equivalent of your ISDN equipment, must have multiple functions built into it to provide for the dedicated path you need for each service you have provisioned on your ISDN line. For example, as you will see later in the order-code table, TABLE 4.3, there are many variations of a BRI circuit. These services require that each BRI line card have two voice ports and three data ports. At most you can be using only three of these, but which three and when is completely random, making it necessary to give you five ports to provide all the options for your service. The same is true for a PRI circuit, where you can have up to 24 voice or data calls, requiring 48 ports for this line card.

One other point about the facilities on the first leg of the call (path 1 in Figure 2.2, also known as the local loop), there are distance and equipment considerations. Typically, any circuit over 18,000 ft. requires some type of repeater device that must be installed by the phone company. For existing analog circuits that are being converted to a digital line, when the analog lines are over 18,000 ft., there will be devices on the line called load coils that must be removed. This contributes to increased lead times to install your circuits, and is an item that is unknown when the order is placed.

BRI — Basic Rate Interface

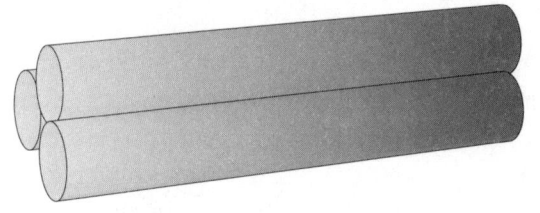

FIGURE 2.3   BRI CIRCUIT.

The BRI circuit is possibly the most powerful version of ISDN available. The circuit is often described as a 2B+D and has a capacity of 144 Kbps. This certainly doesn't compare to the bigger brother, PRI, but the cost of equipment and ease of installation makes BRI a logical choice in many situations. In fact, having a dedicated BRI line for remote computers can give the same feel that a workstation on a network has, allowing you to move work out of the office with less fear of slow response times or expensive hardware.

All ISDN circuits are described in terms of channels, which are a logical way to represent how a cable, with only two wires, is split into more than one type of call or service. The two channel types carry digital transmissions. There is a D channel, or data channel, and B channel, the bearer channel.

EXPLAINING THE TECHNOLOGY

■ D Channel

Let's begin with the D channel, where all of the messaging occurs. It is on this channel that an ISDN service carries all of the intelligent message traffic between a device on your site and the phone company's equipment in the central office. There are a lot of activities that go into setting up and managing a call that are carried in the form of messages on this channel. A more detailed look at the messaging will be covered in a later section of this chapter. Also, the messaging traffic enters the phone network at your central office, where the local phone switch then performs the requested function without creating a dedicated circuit to the location you are calling.

> X.25 is a packet protocol to carry data through a public or private network, with address and error correction information contained in each packet to route the packet through the network.

This D Channel operates at a speed of 16 Kbps, compared to the speed of each B channel, which carries data at a speed of 64 Kbps. The D channel is also capable of carrying data, in the form of X.25 packets, which can be passed to the public packet switching network. Unless the D channel is being used to transmit packets, the D channel will sit idle until it is time to drop the call.

■ B Channel

The two B channels carry the actual ISDN data for your voice or data transmission. You might recall that your voice is packaged into a digital data stream. Your ISDN terminal adapter (TA), the equipment located at your site, performs that conversion and packs it into a series of ISDN packets to be transmitted through the phone network. None of this data will be transmitted until the circuit has been set up via the messaging on the D channel; then the B channel information is passed through the phone company transmission facilities. The B channel data travels untouched, except for the translation of the format to carry the phone company's network messaging.

One of the options you have available with the B channels is the ability to carry packet data, just as the D channel does. Several advantages arise from this capability. The first is having a

64 Kbps packet connection, four times the D channel speed. The second is the ability to allow multiple remote sites access your site through this one connection. This is based on the capabilities of X.25, but is limited by the connection your central office has with the public or private packet networks you need to access. Check with your telephone company to determine if you can implement this feature.

Another great feature of BRI is the ability to have up to eight devices on a single line. Only three activities can be going at once, but through this feature you have the ability to create a miniature phone and data system with eight stations, and only one wire pair to deliver the services to your premises. This capability relies on specific equipment that will be covered in the hardware chapter and again in the application examples.

Now that you understand the basic principle behind the BRI circuit, you'd expect that every ISDN device would talk to each other without a hitch because they are all the same. Sorry, doesn't work that way. In the standards section you will find out that there are different flavors of ISDN. Here's the good news: To deal with this problem, there are standard protocols that package data in such a way that all of the ISDN circuits will move information between any two sites, leaving the messy details of getting two ISDN standards to talk to each in the hands of the telephone people.

PRI — Primary Rate Interface

While the BRI circuit has the flexibility and affordability to outdistance PRI for the number of circuits installed, the PRI is the workhorse of this technology. With a top speed of 2.044 Mbps along with 30 B channels and a single 64 Kbps D channel (up from 16 Kbps on the BRI line), there is a tremendous capacity to haul your data or voice calls.

Much of the initial use of PRI has been for the sophisticated voice services it offers. The advantage of the ISDN technology is

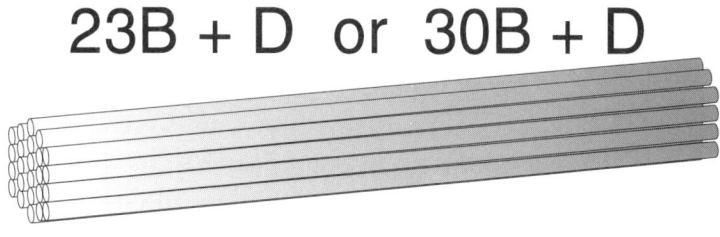

FIGURE 2.4 PRI CIRCUIT.

that any channel can act in any preconfigured way you would like it to, so it is possible to have a channel carry many different types of voice calls that previously required a dedicated line from the phone company.

Just like the BRI line, this flexibility in how an individual channel is used comes from the power of the messaging on the D channel. Instead of 16 Kbps for the D channel, a PRI D channel operates at 64 Kbps like the B channels. This increase in speed helps to handle what should be a more active and dynamic group of circuits.

While the BRI line can be combined with other BRI lines to give you increased bandwidth, each of the lines must use their own D channel to manage the call, so three BRI lines used for a videoconference call would act as independent calls, complete with dialing, call setup and monitoring, and call tear down responsibilities. The PRI line can combine channels with others, and you only have to dedicate a single D channel to control up to 20 PRI circuits or 479 channels. In practice you will not want to put all your eggs in one basket like this; you will back up the D Channel by designating another D channel on a separate PRI circuit to keep your services running in the event of the failure of the primary PRI circuit.

This circuit is delivered on a digital entrance facility, commonly called a T-1 circuit (some times called a DS1) in the United States, Canada, and Japan. The total capacity of that circuit is 1.544 Mbps, and if you do your math, you recognize that this is

The extra bits in the T-1/E-1 circuits are called framing bits.

not 24 channels at 64 Kbps. The difference of 8Kbps is used to format the data into the T-1 circuit format. For ISDN outside the United States, Canada and Japan, the E1 circuit carries the PRI service. E1 is actually 32 64 Kbps channels, where one channel is used to format the data, while a second is used for the D channel, leaving 30 channels for your use. Table 2.1 gives some additional line speeds for the digital circuits that carry ISDN services. The higher rates are used in the Broadband ISDN services.

TABLE 2.1 Comparison of DS versus E channel rates.

DS			E		
Level	Channels	Speed	Level	Channels	Speed
0	1	64 Kbps	0	1	64 Kbps
1	24	1.544 Mbps	1	30	2.048 Mbps
2	96	6.312 Mbps	2	120	8.448 Mbps
3	672	44.736 Mbps	3	480	34.368 Mbps

Another advantage to a PRI line is the ability to make multi-rate calls, where the call can be at (n) times 64 Kbps rates. Some of the most popular rates are H0 at 384 Kbps and H11 at 1.536 Mbps or H12 at 1.920 Mbps, the full capacity of the circuit. TABLE 2.2 gives you a quick review of the various H-channel groups and possible applications at those speeds.

What makes this exciting is the replacement this offers for fixed bandwidth circuits. Today, if you want a 384 Kbps circuit for videoconferencing you would install a circuit to do this, and it would always operate at that speed. If you need only 128 Kbps or 1.536 Mbps later on, then you would order a new circuit. But with multi-rate PRI you can determine the speed at the time of the call, and only pay for the capacity you used, instead of the capacity you provided for and paid to maintain.

You do need to watch out for a critical difference between the use of digital phone lines and the analog ones you use today. Power is furnished by the phone company for most of the single

TABLE 2.2 H channel group applications.

H Channel Group	Rate	Applications
H0	384 Kbps	Data, voice, facsimile, standard broadcast digital audio, compressed videoconferencing
H1	1.5-1.9 Mbps	Standard compressed videoconferencing, private network trunks, data, PBX access
H2	32.8-44.2 Mbps	Fast scanned compressed video, high-speed data
H3	60-70 Mbps	Broadcast video, data, high definition television
H4	135.2 Mbps	Broadcast video, data, high definition television

> This flexibility gives rise to the idea that you can consolidate a number of circuits, spread throughout your facilities, into a single PRI circuit and make use of the dead time these circuits might have during different times of the day.
>
> For example, you might have a few switched 56 circuits being used for transferring data between remote sites and your central computer. And now a couple of your executives have decided it's time to do some telecommuting, and they don't want the slow speeds of a 28.8 Kbps modem. To top that off, you just picked up a headquarters office 2,000 miles away, where everybody loves the weather and nobody wants to travel to see you in the dead of winter, so here come the videoconferencing lines.
>
> Whew! That's a lot of circuits, with most of them sitting idle during the opposite time of the day that the others are working. Your ideal solution would be to reconfigure the bulk of these circuits to ISDN, routing the lines out to each of the areas that need the service; and then balance the work load so that the chance of one function stepping on top of another is minimized. Instead of 10 or 12 circuits, many with expensive monthly costs, you would have a single PRI or a small group of BRIs. Instead of Switched 56, you would have 128 Kbps, reducing the connection times on your data calls.

line analog sets you use, but not for the digital lines. The digital lines require you to provide the power, and if that power goes away, say good-bye to your line and all of the great services it provides. With that in mind, you will need to have some type of UPS equipment to provide power when the lights go out. Besides providing this blackout protection, a UPS will also protect the equipment from possible electrical damage during lightning storms or electrical surges.

> **UPS —
> Uninterruptible
> Power Supply**

Messaging

As I have said, the magic of ISDN is in the messaging capabilities. This simple idea, that devices should communicate with each other, allows the equipment providers to program many useful features into this technology, and those new features can be added to many devices through simple software upgrades.

In Figure 2.5 there is an example of the message process for a simple voice call. You can follow the flow of the call, starting with the initial lifting of the handset, which creates a request for dial tone. If a B channel is available for a voice call, the TA will provide dial tone, and you can dial your number. This number is then sent to the phone company where the request is routed through the SS7 network. If the line of the person you are calling is available, then that phone will start ringing and your TA will receive a message instructing it to give you a ringing signal. Once the call is answered, the central office switches notify each other and your TA that the call is connected, and then the B channel and the dedicated phone network circuit are made available to your voice call. It isn't until that last step that the phone company dedicates any circuits to your call; everything is done in the D channel and the SS7 network.

After a call is completed, when either party hangs up, again a message is sent through the SS7 network and the D channel to release the voice circuit. The number of messages sent are more numerous than this simplified overview.

You now know that your voice is a data packet within the phone network. And controlling delivery of that data packet is an easy task for the equipment. One of the voice features, multiple lines, is provided by controlling the flow of voice packets. To

EXPLAINING THE TECHNOLOGY

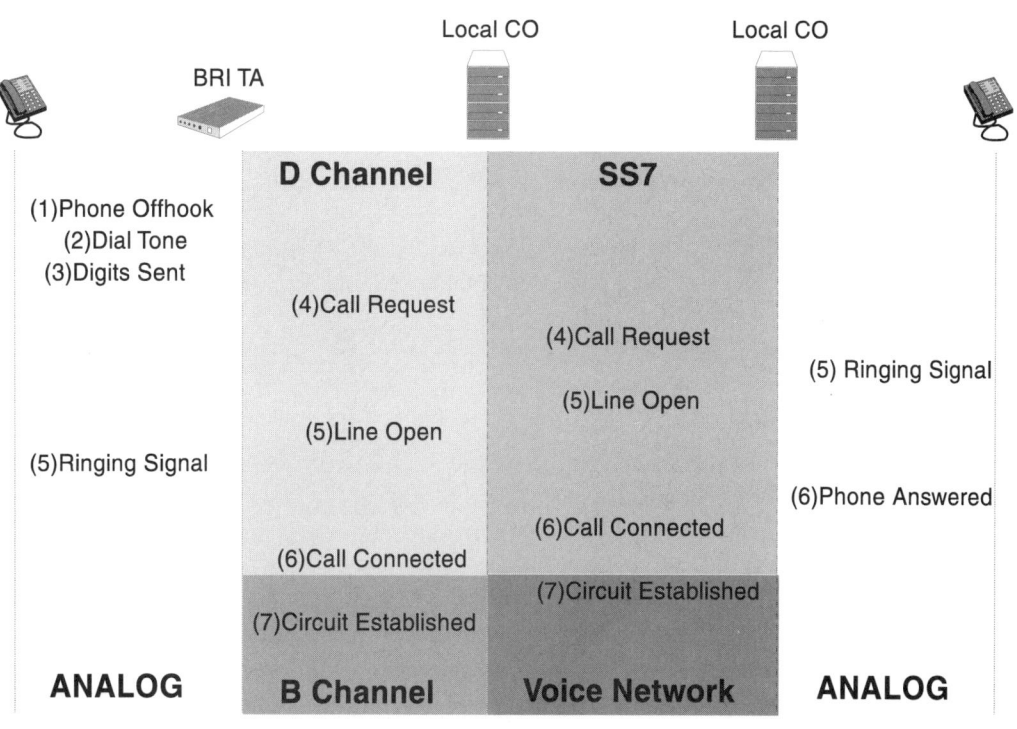

FIGURE 2.5 MESSAGING FLOW FOR A BRI VOICE TELEPHONE CALL.

place a call on hold, you only have to send a message (the hold button pressed) to the phone switch, which stops sending the packets. To enable a second call, the switch just routes the second call's packets, while the first call sits waiting at the switch. Switching between the two calls can be as simple as selecting the button associated with the call. It is even possible for analog phones to use these features, although it is not as easy as an ISDN phone.

In the applications chapters you will learn more about specific features of BRI and PRI to increase your knowledge of the technology. But for our strategic decisions and design needs, you're almost armed and dangerous enough to begin coming up with all sorts of things you can do with this technology. The next subject to cover is the standards that ISDN is delivered through.

31

■ CURRENT AND EMERGING ISDN STANDARDS

It is because of the intense effort of developing mutually agreeable standards that our communications infrastructure has grown to be so robust. While the developers of the technology rush ever forward to create new technologies, the standards committees are there to ensure that this critical resource will remain open and available to everyone, almost regardless of their technology to connect into the global network.

Standards Bodies

There are a number of mechanisms to create a global sense of interoperability, and these are accomplished through the standards bodies of the world. International standards are set, and then enough room is left in them to let the individual countries create a set of national standards to meet their own needs.

Through the process of the standards bodies, shown in Figure 2.6, the shape of telecommunications services are defined. As a multilevel committees process, it is a wonder how anything gets done at all, but done it does get and today's robust telecommunications infrastructure is a strong testament that it works to provide us with a global ability to communicate.

The International Telecommunications Union (ITU) is a treaty organization of the United Nations and was known as the CCITT (International Telegraph and Telephone Consultative Committee) before a recent name change. The history of the ITU dates to 1865 and has contributed greatly to the immense success of the telecommunications industry. Membership in this union is composed of the governments of the world, not individual industries or companies.

Within the United States, the Telecommunications Industry Association (TIA) is responsible for the standards of the equipment connecting to the U.S. Telecommunications infrastructure. Unlike the ITU, this group is composed of the various members of

Explaining the Technology

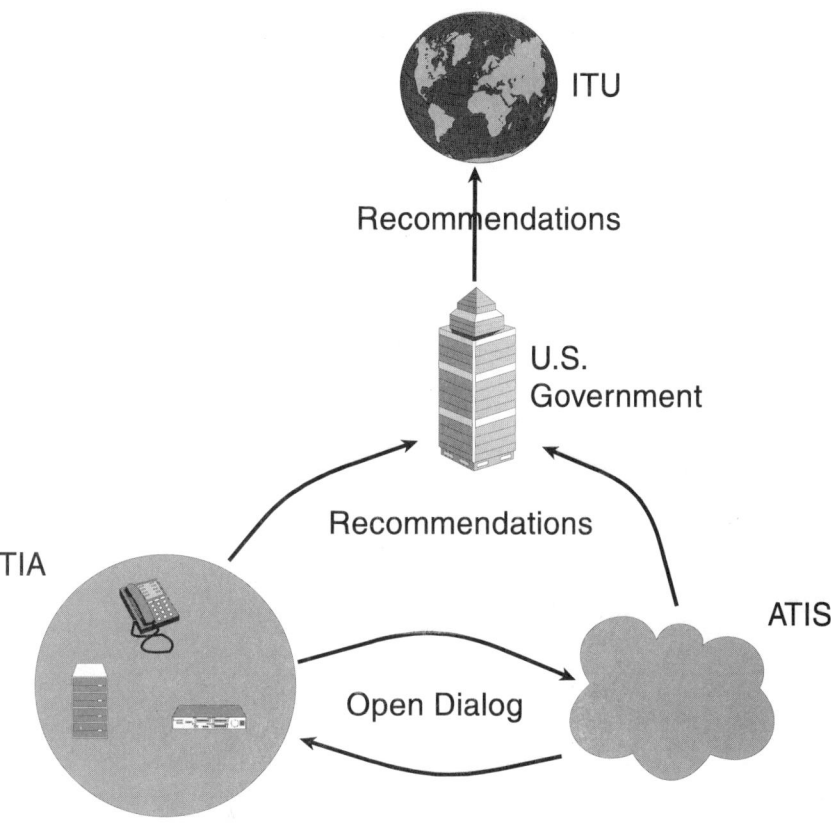

Figure 2.6 Standards creation flow.

the industry, working to provide the connectivity you want, while preserving their competitive advantages with proprietary designs. This is a fine balance that companies must walk in the marketplace, and it places challenges on you to decide what solutions do meet the standards, or are positioned to become the next standard, and yet take advantage of the full range of the technology.

> **Carrier —** These are companies like AT&T, NYNEX, etc.

While the TIA covers the equipment, the Alliance for Telecommunications Industry Solutions (ATIS) establishes the standards for the network itself. This area covers how your ISDN call will be transported through the network, across the many different transport technologies the carriers use.

National ISDN and Services

Perhaps one of the greatest barriers to the emergence of ISDN in the past was the lack of standards that the industry would follow to insure interoperability. It wasn't until 1991 that the National ISDN 1 standard was agreed to and products emerged that communicated with each other. Since that time, in the United States and Canada there has been three levels of ISDN proposed. These are labeled National ISDN-1, -2, -3. Right now the National ISDN-1 standard is rolling out and should be available in most areas, with full coverage expected in late 1996 or early 1997. National ISDN-2 is in the wings waiting for upgrades to many of the central office switches, expecting to begin rollouts in 1997 and with National ISDN- 3 to follow that.

It helps to understand what each standard does for you, but I won't spend much time here explaining all of the features made available through each standard. The detailed services and features of a product will be defined within the scope of what level of National ISDN it supports. For more details on these standards and features look in Appendix D or on the CD-ROM in the NIUF guide.

When you are designing your ISDN application knowing which standards, and what services the carrier will make available will influence the choices you make in the short term — limiting your features and forcing you to look at upgrades to your equipment down the road. So remember to think forward on your needs, and then make sure that your vendor and consulting group can tell you how your solution will migrate forward.

■ National ISDN-1

National ISDN-1 is the core standard to bring all of the custom implementations of ISDN together. Through this standard there are a minimum number of services, such as: basic voice and data calls, call hold, conference/transfer, call forwarding, and other features that provide you with all of the services that have in

EXPLAINING THE TECHNOLOGY

> **Key System is a small phone system with limited capabilities.**
>
> **PBX is a Private Branch Exchange, with many of the same features that a central office might have.**

the past been delivered from a key system or PBX. Within the data capabilities, National ISDN-1 allows for D- and B-channel packets and "on-demand" B-channel packets. Through this standard it is possible to initiate an ISDN call from anywhere in the network and talk with any device meeting these standards at the other end.

It's important to know that AT&T has a custom implementation of ISDN on their 5ESS switches. At first encounter, the word custom goes with proprietary, a deadly word in an open, standards oriented world. However, the 5ESS implementation encompasses many of the National ISDN services, and AT&T is working on additional releases of their software to bring their product closer to the standard, without giving up the competitive features.

Many of the terminal adaptors, the phone switches, and other hardware that are being manufactured for this level of the standard are still made to work with the central office switches that haven't been upgraded yet. The next level of National ISDN is coming and with it comes new features and hardware requirements. For equipment to undergo transition to the next level, it will have to be programmable. Many are, so this should be less of a problem, but some of your equipment may not make the transition.

■ National ISDN-2

Perhaps the most powerful feature of the next level of ISDN standards is the provision for the Automatic Terminal Setup. Simply put, this will be the closest thing we have seen in terms of plug and play networking. As you will find out in Chapter 4, there are all sorts of details that have to be resolved to install ISDN today. Not quite so in National ISDN-2.

The industry has come together to define a standard set of order codes to ease the provisioning of ISDN lines at the BRI level. With those standards in place, the equipment manufacturers are now committing to put those codes into their equipment, as a way of telling a phone switch, *"Hi, I'm a CiscoPro 753 Router and I wanted to let you know who I am. Also, I think I'm expected, could you tell about where I am and how we're going to work together?"*

This little conversation sounds like a fantasy, but when these devices and switch capabilities are available, some time in 1997, you will be able to just plug your ISDN equipment into the circuit, and in a few minutes, begin to configure your application without the worry of all of the addressing and protocol details that exist today.

Other features of the this level of the standard include:

- Switch DS1 and Switched Fractional DS1 (SWF-DS1) services.
- SWF-DS1 for packet-switched data.
- On-demand B-channel packet-switched data on BRI.
- Enhanced call control features on PRI, including per-call services.
- Increased billing capabilities to support usage-based services.
- Support for eight devices on a single BRI.
- Consistent implementation of BRI features, allowing you to move equipment between central office switches without any noticeable difference.
- Increased Electronic Key Telephone Service (EKTS) features including: multiple directory numbers per terminal, analog phone in the group, multiple directory number appearances and call appearance call handling, hold/retrieve, bridging, intercom calling, multiline hunt group, and special ringing features.

The full set of features is extensive, but this list shows that your ISDN phone is actually a system all by itself. These extensions to National ISDN give you the ability to handle phone calls in the manner you would in a large corporate office, without all of the expense.

■ National ISDN-3

While many of the features detailed on National ISDN-3 are going to support enhanced maintenance functions, there are a number of new capabilities that will be released to the users. The original expectation was that 1995 and 1996 would be the years that the industry rolled out this level of ISDN standard. Unfortunately, everybody back in 1992 were overly optimistic on the time frames, but fortunately, not on the technology.

Some key features in this standard will be:

- Noninitializing terminals — specific function devices not requiring all of the configuration and support of the TAs today.
- Directory number shared over multiple devices.
- Music on hold.
- Calling name delivery for all ISDN calls.
- ISDN inspect — you can see what features you have by asking the phone switch.
- Call by call access to the long distance carriers.
- Default services for terminals — allows for 911, 611, or other basic calls on uninitialized terminals.
- PCS internetworking — the Personal Communications System will work with ISDN to provide features like: automatic link transfer, three-way calling, calling number identification, and call forwarding.
- Access to frame relay — through a frame handler, you can access a permanent virtual circuit (PVC) in a frame relay network at 64 Kbps.
- Multipoint video/data conferencing — with a multipoint control unit you can link multiple sites into a single conferencing session.

> **Frame Relay —** a data packet network that is replacing X.25.
>
> **PVC —** a logical connection between two sites in an X.25 or Frame Relay network.

- Modem internetworking — through a device in the phone network, calls can be converted between analog and ISDN formats, allowing for seamless access to remote equipment regardless of equipment type.

- Multicasting — through a server device in the phone network, you can broadcast a single data packet to many locations.

Many of these software features are still being developed, with a phased rollout over the next few years. What this shows us is that ISDN has the ability to create a custom phone system environment anywhere. Small offices, home-based businesses and telecommuters, even folks that just want to have some phone services that are a bit smarter than what they have today, will see these capabilities become real with ISDN over the next few years.

The planned integration of ISDN with the PCS wireless systems that will be rolling out in 1996 and beyond, gives you some great capabilities in phone calling. With the ability to switch between call points, you could have a call on your PCS phone move over to your land line at home, getting you off the more expensive airwaves and back to a lower call rate. It will work the other way, you could start a call on a land line and carry it with you through the PCS phone.

Protocols

The actual capability of talking to one another through an ISDN link actually occurs at a higher level than the equipment, it occurs in the protocols. To help you put this in perspective, I will briefly review the OSI model and where the various pieces of the ISDN puzzle fit.

OSI — Open Systems Interconnection from the International Organization for Standardization.

OSI is a seven-layer model that describes functionality to manage the process of networking. Core to this functionality is the idea that no layer needs to understand more than the interface to the layer above and below it. And at both ends of the line, each layer talks only with the layers around it, and the messages passed through the network are intended only for the same layer at the other end.

In Figure 2.7 you find the basic OSI model that describes seven layers of functionality in a networked environment. As you work your way down from layer 7, applications, you move farther away from the user-interface where the world might still look friendly and inviting, and you enter a world of acronyms and functions that are best left in the hands of highly skilled technical network folks.

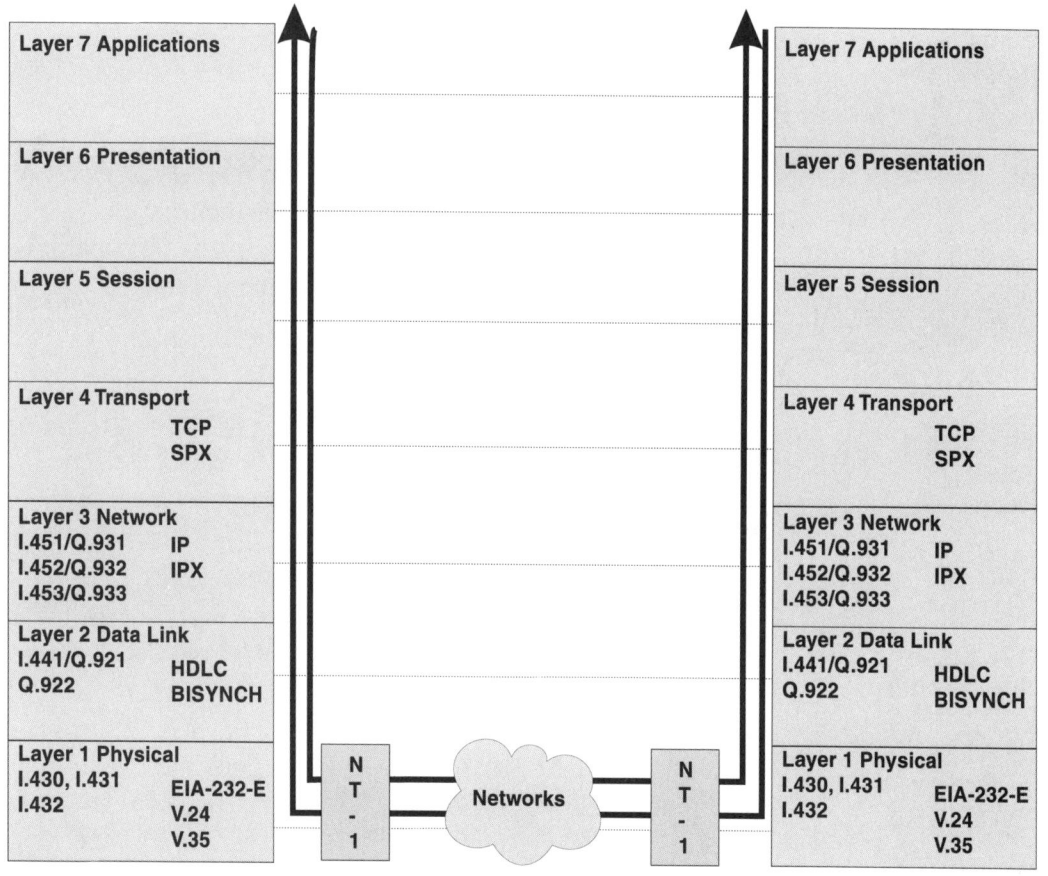

FIGURE 2.7   INTERNATIONAL ORGANIZATION FOR STANDARDIZATION OSI MODEL.

The intention of this model is to allow designers to focus their efforts at particular layers in the model. Each layer has to communicate with three entities: the layer above it, the layer below it, and the same layer at the other end. Through this model, all of the controls to support a robust and reliable network have emerged. This has also simplified a lot of functions, because you can switch out underlying layers without affecting the layers above, and vice versa. This supports the implementation of new transport mechanisms without impairing the large base of applications using a given technology.

A good example of this switching of a layer is the ability to move ISDN to a wireless environment. In this situation, either through satellite links or the coming PCS spectrum, you can transmit ISDN circuits because while the physical layer has been changed, from a copper wire to an electromagnetic wave, layer 2 in the model doesn't see this change, as long as the physical layer maintains the same interface to layer 2 that the old interface did.

You can see the benefit from this structure; it allows us to isolate devices and technologies from the data to be transported. For example, in many areas the supported bit rate on ISDN is less than the capacity of the circuit, and this gives rise to a problem that this model solves. If you are running at a rate of only 19.2 Kbps across a 64 Kbps link, what happens to the rest of the bits?

> Encapsulated — a process where an element, in this case, a data packet is surrounded by additional information to make a new packet in the format required by ISDN.

The V.110 protocol standard handles this problem. With this protocol, the data from your computer or other data communications device, formatted for a 19.2 Kbps transmission, is encapsulated into the 64 Kbps packet that the ISDN circuit is expecting. Then at the other end, V.110 reduces the packet back to the original 19.2 Kbps worth of data. This protocol is prevalent in Europe, while in the United States it is V.120, a 56 Kbps protocol, that is used.

Other standards will address different application areas. For example, with the rollout of the H.320 videoconferencing stan-

Explaining the Technology

dard, many otherwise proprietary video systems can now communicate with each other. This has enabled large-scale conference room systems to transmit to other conference room and desktop systems, without caring about the other end. This simple feature has opened up the videoconferencing market to hundreds of thousands of new users, especially with the advances in video compression technology and the speed of a BRI line.

As standards are used, they will be introduced in greater detail in the application chapters. Right now you should understand where they fit in and what some of them are by name.

To finish the chapter, let's take a look at the higher speed ISDN technologies. These transport protocols are part of the ISDN family, although they have taken on a a level of recognition all their own.

■ Broadband ISDN — Bandwidth on Demand

SMDS — Switched Multimegabit Data Services.

ATM — Asynchronous Transport Mode.

The are a great many options in our future quest for higher speed. In the previous chapter I discussed the alternative technologies (cable modems and DSL) that will bridge the bandwidth gap between the present ISDN offerings and the technologies that encompass SMDS, ATM, and their underlying transport technologies. Between these services they offer speeds starting at 1.5 Mbps with SMDS and scale to 155 Mbps and beyond with ATM.

SMDS — Switched Multimegabit Data Service

SMDS is an interesting service offering because it brings with it the ability to get high speed switched data services from 1.5 Mbps up to 45 Mbps. This service in not a dedicated data-circuit service as BRI and PRI are today. Instead, data packets transmitted into an SMDS network containing data and address information, allowing packets to enter the SMDS network, and to travel with other packets, sharing bandwidth in the most efficient way possible.

This service is designed for applications where there is a need for large bursts of information over reasonably short periods of time. To use this service, you purchase an access path to the SMDS network, and then you simply connect to the network as needed and transmit your data to another location. Applications on this type of network could include near real-time video editing, where you could set up a collaboration between people across large distances, and they could all view and work with the material in a real-time setting.

ATM — Asynchronous Transfer Mode

Behind this simple acronym, ATM, is a service that holds most of the promise for future bandwidth in our communications infrastructure. Its power comes from the simplicity of the transport protocol and the layer of the OSI model in which this technology operates.

> Cell —
> 48 bytes of data and 5 bytes of addressing information.
>
> Asynchronous — data travels at irregular intervals.

ATM is a physical layer protocol, with a very simple 53-byte cell format. It is the simplicity and small size of this cell, combined with the asynchronous nature of the transport that gives ATM the ability to travel at any speed we are capable of providing in the hardware.

Sitting in the middle of Layer 1 (Figure 2.7) of the OSI model, ATM has split the layer into three pieces. Above ATM is an adaptation layer that talks with the level two protocols, while beneath ATM is the physical transport medium. Again, the modularity of the structure allows us to move ATM to many different transport systems.

In the wide area networks you will see speeds of 155 Mbps associated with ATM, with demonstrations of 622 Mbps going on now in the labs. In modeling, this technology shows no practical limit to its ability to route traffic. There is no error correction in this protocol and, basically, no need for it. Error rates in our digital networks are so small, that it is much less expensive and that much more productive to let the higher level protocols deal with the errors.

But this upward mobility is not the only place for ATM to go. Networks are being deployed today using ATM as the backbone, at speeds of 25 Mbps. Additional work is being done to move ATM all the way down to the 128 Kbps speed in the wireless network environment. As these networks become more prevalent, a new class of routing devices will emerge. If our data are formatted in this ATM structure, than linking to the public ATM network will become much simpler, and much more effective. Add to that the fact that our voice traffic can also be reduced to this structure, and now you can move voice and data together, down one pipe, letting packets head off wherever they need to.

If you carry this idea forward in time, perhaps as soon as 10 years from now, tremendous things begin to emerge. Begin to think of your phone systems, data routers, bridges, and modems as communications gateways. In this frame of mind, what prevents us from using one simple gateway, if all of the devices in our communications network talk the same language, ATM, or go through devices that reformat their packets to ATM. In that case, the routing of calls, data, and video become easier because you can notify the network what physical address you want your packets shipped to, and the network will automatically reroute your traffic.

In this vision of the future, even more of the software and computer technology used in today's telephone systems and computer networks would be merged, giving us the ultimate personal information devices the industry has been prophesying for so many years. I'm not predicting the time frame for this, but I am saying that the networking infrastructures required to make this happen are finally within sight, and that makes it only a matter of reasonable time before this will happen.

■ What's Next

Now that you have a rosy glimpse of the future, let's get back to the business of today. In the next chapter the basics of the hardware and ISDN lines will be explained, giving you the knowledge to understand the framework of the solutions you will see later on.

CHAPTER 3

Explaining the Hardware

In this chapter you will learn about the kinds of physical equipment you need to implement an ISDN solution. The tour will begin with the basic wiring, explaining the types of wire and the various wiring configurations you might find installed today. Then on to a review of the telephone companies and their networks, the various connection devices, telephones, computers, videoconferencing equipment, and other innovative ideas.

■ WIRING

A wire is a wire is a wire, right? Not anymore, nor has it been for many years. In fact, in many instances, a wire is just an analogy to what is actually happening, giving us a familiar framework to understand what the technology is trying to do. On the other hand, the real wire is still the way your ISDN service is delivered, along with your traditional phone service and now your computer network.

For most of the applications you will implement today, the wire will truly be pairs of copper wires dedicated to your ISDN circuit. But not all pairs are made equal, and in many instances, you will find it advantageous to install new wire, and that will lead you to choices I will guide through.

Wire Categories

Basic telephone service has been delivered on what is termed Category 3 cable for many years. The Category definition is an EIA/TIA standard that defines the minimum performance level of the wire. In the case of phone service, that minimum is to carry voice-grade signals up to about 18,000 feet without repeater devices. In computer terms, this type of cable is capable of carrying data traffic at speeds up to 100 Mbps, over a much shorter distance of hundreds of meters, certainly much faster than the present 128 Kbps that BRI provides, or the 1.536/2.048 Mbps of PRI channels. So it seems that Category 3 is all you need, right?

No, not if you consider your entire organization and all of the technology that might go that wire. If your chief financial officer wants to control both short- and long-term costs, your choice of wire could affect your having to rewire the facility in just a few years. Your objective should be to spend enough money right now to keep the wire in the wall for more than a few years, and to allow your technology to expand when you need it to.

Wire Types

To finish the basics for wire, TABLE 3.1 has a brief description of each wire category. Overall, the Category 5 wire is the best choice when installing new wire, because it can carry signals at speeds much higher than the technology permits today. For example, where Category 3 cable has reached a practical limit of 100 Mbps, the Category 5 cable operates at 155 Mbps today and is working in the labs at 622 Mbps, while the physicists tell us that 980Mbps is the theoretical limit. The real point — wiring a facility is expensive, time consuming, and disruptive, so make your best choice for the next twenty years and pay a little more up front.

TABLE 3.1 Wire category characteristics.

Category	Characteristics
2	This is an old wire type, two pairs typically, that was used in many homes and businesses. OK for the original voice services, not usable today.
3	This is the minimum level of wire installed today. It is good for voice and data up to 100 Mbps.
4	An intermediate wire category that handles applications up to 20 Mbps for data, plus voice.
5	At the top of the wire types, this cabling is capable of all voice applications, plus data in excess of 622 Mbps.

Wire Closet — a dedicated area that you bring all of your cabling to before connecting it to your phone or computer systems.

The primary difference between these cable categories is the amount of twist in the cable. Where the Category 3 cable has a twist every six inches, the Category 5 cable will twist every one and a half inches. This twist structure contributes to the ability of the cable to carry electrical signals, and the tighter the twist, the more effective the cable. This twist characteristic extends to the connector blocks, the wall plates, and the jumper cables that you might install in the wire closet. If all of the equipment is not installed properly, or there is a mix and match of categories of cable, then your entire wire system could contribute errors into your data or phone networks. So be sure your cable installers are certified Category 5 technicians, make sure they test your wiring after installation to verify that it still meets Category 5 specifications, and get their warranty on the installation and product. Most of the cable systems are guaranteed three years when installed properly.

There has been a long-standing debate about the use of shielded versus unshielded pairs. This debate stems from the idea that the cable can act as a shield or an antenna. The fact is, it does both and the design of the cable contributes to how little or how much it does so. Shielded cables have been the rage for many years, because the shield, a metallic cover around the entire cable, was thought to reflect outside noise, and by being grounded, it would drain off any noise from inside the cable. This worked when everything was installed correctly. Problem is, even if it was installed right, with a simple break in the shield or its connections, the entire cable turned into an antenna, making matters worse. Unshielded cable doesn't have this pitfall. The design of the cable tries to minimize the impact of the antenna effect, and overall, the recommendation today is to always install unshielded twisted pairs (UTP) unless instructed otherwise.

Number of Wires and Connectors

Cable that is typically installed today will have eight wires, or four pairs, to carry both communications signals and power. How you use this wire is important, because your design can take you from using only one pair to needing three pairs to implement your application. As a simple primer, Figure 3.1 shows the typical RJ connectors and which pins are used for each wire pair.

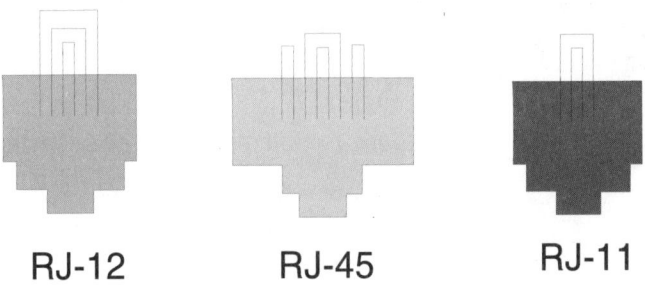

FIGURE 3.1 RJ WIRING CONNECTOR CABLE PAIRS.

ISDN enters your facility using only two wires, one pair, and is broken into four signal wires by the NT1 (a device you will learn about in the next section) to be delivered to the ISDN devices. So if you are only using four wires in a Category 5 cable, and you have four wires left over, what could you do? To start, you can consider running your LAN down this cable, since it only needs two pairs to operate. Or if you are putting in the ISDN line as a supplement to your existing digital phone system,

Explaining the Hardware

then you should only have one pair used, leaving three pairs for the ISDN circuit.

As you may recall, ISDN circuits do not carry any power, unlike your analog phone service, so you must provide power to the equipment. This power can be delivered on a single pair of wires along with the ISDN pairs or from the equipment you are connecting. Each of the designs you see in the book will show you how many pairs you will need to make the application work, and what pairs will be left for power or other applications.

Now, why wouldn't you want to share the cable with another service. From a practical point of view, if you are wiring from the ground up, put in a cable for each function you need to support. That means at least a data cable and a voice cable to every location. The advantage to this is that there are spare wire pairs that are available in the cable in case a problem arises later on. But don't plan on the spare pair coming from the same cable, typically, cable doesn't break as an individual wire pair, it breaks as a cable. Two cables with spare wires will allow you to fall back to a single cable, deferring another cable pull until it is convenient.

Wire Closets

In all facilities, the wiring begins in a central location, called a wire closet. In larger facilities, there may be multiple closets, arranged as shown in Figure 3.2. The wire closet provides the space and equipment to organize your wiring for effective use and troubleshooting.

Punch down blocks, patch panels, and patch cords are the equipment pieces to connect and organize your wiring (Figure 3.3). Punch down blocks are used to cross-connect wire from a source, like the phone company's wire entering your building, and the destination, your phone system. This block has connectors to which the cable is connected by literally punching the wire into a connector, creating a connection between the wire and the pair of connectors on the block.

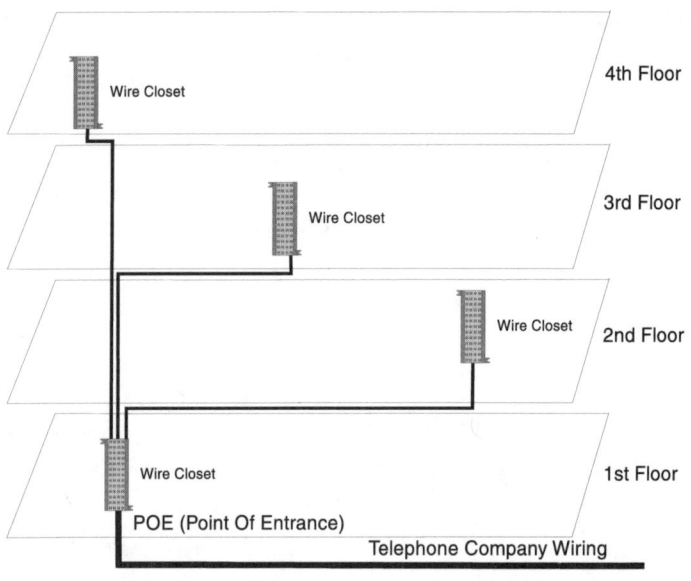

FIGURE 3.2 TYPICAL CONFIGURATION OF WIRE CLOSETS FOR DISTRIBUTING WIRE WITHIN A BUILDING.

The introduction of a patch panel, which goes between these two points, allows you to switch cable pairs by simply plugging in a patch cord between the wire pairs you want to connect. In a matter of seconds it is possible to isolate complete sections of cable and determine more precisely where problems may exist.

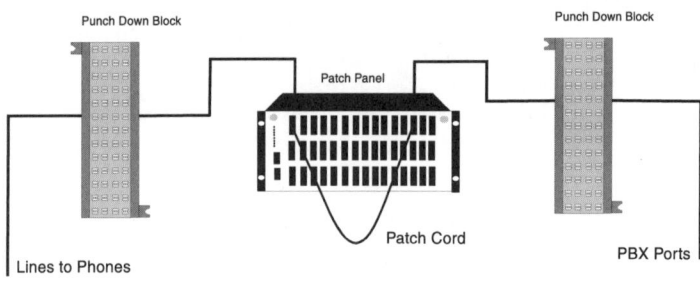

FIGURE 3.3 PUNCH BLOCK AND PATCH PANEL CONCEPT.

Backboards and wire cabinets are the places where these components are mounted. It is very typical that the phone company will put their wire onto punch down blocks mounted on a backboard. The backboard is just a thick piece of plywood that is secured to the wall in your POE (Point Of Entrance, the location in your facility where the phone company places their wire and turns the responsibility over to you to maintain the wire) wire closet. Typically, all of the wire is then distributed through additional blocks located on this board to the various pieces of equipment or locations.

With the proliferation of computer networking equipment, the cabinets traditionally used to house test equipment began appearing as an alternative to the backboards for mounting and maintaining wiring systems. Today, you will find complete solutions built from components mounted (racked) in these cabinets. This has the advantage that if additional equipment needs to be added, relocated, or replaced, it is easier to do so from a rack than from a permanently mounted component on the wall. Another advantage to the cabinets is access to these components from the front and rear, making support easier.

Reusing Wire

But if you are trying to make do with your present wiring, check a few things out first. First, make sure you have at least Category 3 wire. Second, verify that there are enough pairs left to support your ISDN application. Third, test the wire by connecting an existing circuit to it. Also, if you are going to use the circuit for data, try a 28.8 Kbps modem on the circuit to see if you have any obvious problems with the quality of the wire. If you're not up for testing this yourself, have your cabling folks do an end-to-end test of the line.

For those of you in home-based offices, a few more things need to be checked. First, there may not be enough wire pairs coming to your home to add your ISDN line. This will affect your installation cost only if the phone company needs to bring new wire to you. They will pay for the installation to the edge of your property; you get to pick up the tab after that. Second, inside the house, you may only have two pairs of wire, limiting the reuse of this wire. In Chapter 7, Figure 7.4 for example, you will find some creative ways of reusing this wire.

■ NETWORK TERMINATORS

Having the right cable is important, but you don't get to just plug your equipment into the line when it arrives, not without having a required device between the phone company facilities and your equipment. This equipment is required by law to protect the phone company equipment from damage in the event of a failure of your equipment. This special interface is known as a network terminator, NT, with two types available, one for the U-interface point (NT1), and one for the T-interface point (NT2).

The rules governing these devices differ in the United States, where the customer is required to purchase both the ISDN and the NT device. In other countries, this device is provided by the telephone company. If your application is international, be aware that equipment choices for the United States may not be usable in another country, because of power, NT, or regulatory issues. Always check with the equipment manufacturer to determine if their equipment will work outside your country.

Interface Points

If you look at Figure 3.4 you will see the interconnection points for your ISDN circuit into the network. The three interface points, U, T, and S control your ability to add equipment onto the circuit. The U-interface point performs the initial switching from two to four wires, and only one device with a U interface may be present on the circuit. But wait a minute, you heard me say you could have up to eight devices on a single BRI line, what happened? That's where the other interface points come in.

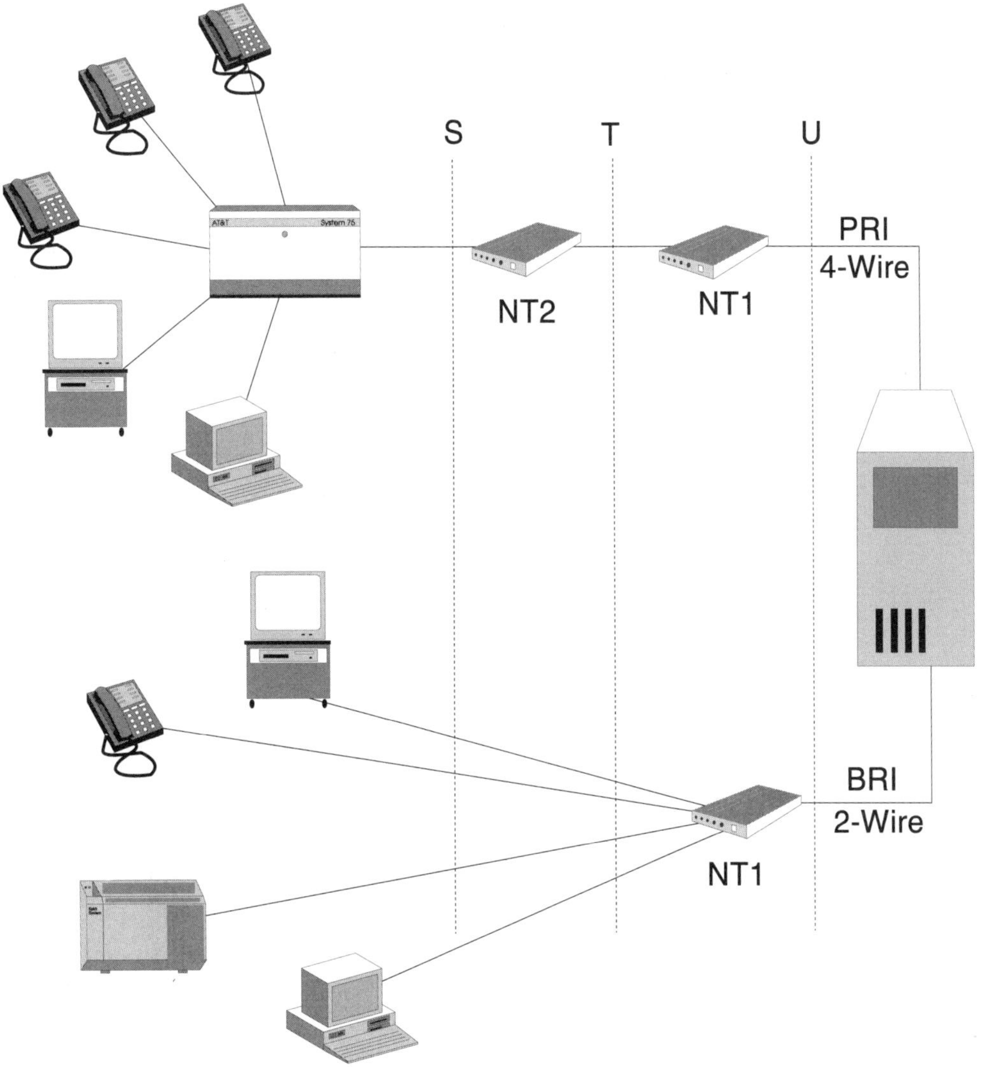

FIGURE 3.4 ISDN INTERFACES TO THE PUBLIC NETWORK.

The T interface, also at times combined with the S and referred to as the S/T, can be used to connect other equipment, provided they have an S/T interface, and not a U interface. Another requirement must be met to successfully run more than one ISDN device on this line, the NT1 must have a passive bus.

The passive bus is a more complex installation, because this approach requires that you balance the total amount of resistance, measured in ohms and limited to 100 ohms for all devices connected. This is not an issue for the PRI circuits, because they are not shared in this fashion.

In the United States many of the ISDN devices are being manufactured with the NT1 built-in. These devices connect to the network at the U interface, and preclude you placing any other ISDN devices on that line, unless the equipment has an S/T interface. The greatest flexibility in the use of the ISDN line comes from the use of the S/T interface, so watch out for this when you are evaluating equipment.

Network Termination Devices

The NT1 is the primary interface to the public network. This device will convert your ISDN circuit from a simple two-wire configuration to the four wires used by all of your equipment. This change from two to four wires results from the technology being used to transmit the ISDN signals across large distances is inappropriate for the shorter cable lengths at your site. The NT2 is used only for the PRI circuit, providing the proper management for this larger circuit.

As you can see from Figure 3.4, there are a number of devices you could attach to your ISDN line. On the BRI side there is a videoconferencing system, with an ISDN telephone, a key system using an analog port from the circuit, and a computer with an ISDN modem. With an NT1 using a passive bus, these devices can share the line quite effectively, and with the ability to assign directory numbers to each device, calls would be automatically responded to by the correct device as needed.

For the PRI circuit, it is possible to now use the PBX as a distribution device for ISDN services. Not only can you route your normal PBX calls down this pipe, but you can actually distribute a BRI circuit from the PBX, allowing you to connect in any ISDN terminal adapter to your PBX, and again share your circuits effectively.

■ THE ISDN EQUIPMENT — PHONES, COMPUTERS, AND VIDEO

We're still at a point where no useful hardware has been connected to the line to let you get anything done. Now it is time to explain what types of equipment are used to interface your actual application to the lines.

Telephone Systems

Now we are going to look at the phone systems you have at your facility. This might be a PBX or Key System, but I will refer to it as just the phone system. In many of the phone systems available today, you can purchase a trunk card that will provide you the connection to either a BRI or PRI circuit. This card will install in the system and then distribute the calls through the internal architecture of that system. Until recently, these systems did not deliver ISDN out the port (phone set) side of a phone system, but that has changed, with some manufacturers now including this capability.

> Trunk — the phone line supplied to you from the telephone company.

■ PBX or Key Systems

You typically have a single card for each ISDN line you are interfacing. One of the reasons that PRI is the preferred choice for ISDN when you have more than about twelve lines being installed in your PBX, it only takes one slot in your equipment. Some manufacturers supply PRI cards that are able to work as T-1 lines (remember, PRI is actually delivered on a T-1) so they can be used in either mode of operation. This is an important feature if you are in an area where PRI is not yet available, but you want to minimize your investment to upgrade later.

A quick note, if you need more than one PRI line from the same carrier, you can install a second interface card. It isn't necessary to have another D channel for this card, you can use the D channel from the first card to manage the circuit, giving you a full 24 channels on the second through twentieth cards. The total number of PRI circuits that can be run on a single D channel is dependent on the equipment you purchase, the service offered by the phone company, and the level of safety you want to build into your system.

■ *Inverse Multiplexor*

If your phone system equipment isn't capable of allocating the channels of your PRI circuit in the way you need, then an inverse multiplexor may be the proper piece of equipment to place into your design. The inverse multiplexor device provides two key abilities in a PRI or T-1 circuit. The inverse multiplexor function allows you to combine individual channels into one larger channel. An example of this function is videoconferencing at 384 Kbps. This is the aggregation of six 64 Kbps channels, which appear to the video equipment as one channel, and to the telephone company equipment as six channels.

If you do need to share the PRI interface with some other devices, like a computer LAN remote access device or videoconferencing equipment, and your PBX or key system can't distribute BRI on the port side, then you can use the drop and insert functionality of an inverse multiplexor. This function receives the PRI line and then breaks it into fixed pieces for each service you need. As such, some PRI channels would be set aside for videoconference, while another might support your remote access needs, and the balance of the lines would be delivered to your PBX.

■ *Telephones*

You can have ISDN or analog phones that directly connect to the NT1 for BRI circuits, and these devices will operate like a phone set on a PBX because of the services available on the ISDN line. Also, remember that to have multiple ISDN phones or other ISDN devices on a single NT1, it's required that the NT1 have a passive bus to allow the individual devices to work together to manage the lines.

Computer Systems

There are a number of devices, internal to the computer, externally attached through serial ports, or connected to the LAN through an ethernet connection. These devices will provide a wide range of services, from data-only to complete fax, modem, ISDN data, packet-data, voice mail, and telephone services. The limit may only be your imagination and your budget.

■ Terminal Adapters

Here we start with the simple terminal adapters, TAs, which are sometimes referred to as ISDN modems. In more precise terms, a TA is any device that provides the translation from one communication technology to the ISDN format. The IBM 7845 NT1 has an S/T port and an analog port. The device serves as a terminal adapter, because it converts standard analog phone services to the ISDN format and vice versa.

Moving up from the NT1 devices with analog ports, today you have terminal adapters with multiple analog ports to give you voice services, plus a serial port for your computer data applications. TAs may have analog fax modems built in, some are now arriving with complete fax, phone, voice-mail, and data capabilities, all on one card. These TAs are available as internal and external devices. The internal devices are installed in computer systems and may be more difficult to get going, but they provide other advantages covered later.

Warning, if you put a terminal adapter into your computer, and you plan on using this to supply your voice lines, it will be necessary to leave your computer on 24 hours a day. It is possible that you will lose the use of your ISDN line because of a failure of the operating system, especially if the failure locks up your PC.

The greatest speed advantage comes from having the terminal adapter inside the computer. A quick lesson in computer port speeds will help you understand why. In the PC world, there are serial interface ports that control data communications with modems and other devices. In recent years the top speed of these ports has been 115,200 bps, using a 16550 UART chip (check your system documentation to determine if you have this chip). Now, when you are transferring information through a computer serial port, the byte is turned into 8 bits, and then there is a start and stop bit added to it, making that single byte take ten bits of transfer capacity. That means that only eight out of every ten bits is your data, so you only get 80 percent of the speed of the serial port, or 92,160. Still with me? This means that while your ISDN line can give you 128,000 bps, your computer serial port limits you to 92,160. Here's the sad part, when the ISDN adapter gets all those start and stop bits, it throws them away on the sending end, and then puts them back on the receiving end if it must send them out another serial port.

> Macintosh Users — your serial port may be limited to 57,200 bps, check your system documentation.

One other note about external TAs and the total speed. There are other serial ports that can deliver greater than 115.2 Kbps. In the United States, these adapters are priced over $200, while in Germany they are readily available for $30 to $50 U.S. This still doesn't solve the problem. The TA is probably still using the 16550 UART, so no matter how fast you can get data to the device, it won't take them any faster than 115.2 Kbps.

On the other hand, internal ISDN adapters don't have this limitation. They talk with your computer through the backplane and can give you the full 128,000 bps that the ISDN line is capable of providing. One small catch here, not all cards have software drivers, the required piece to make this solution work. And if a driver exists for one operating system, it may not for another — so do your homework before you commit to a solution.

If you have a choice, and speed is your primary need, go with an internal TA. If other features, which may not be available on an

Explaining the Hardware

internal TA, are more important, then suffer the lower speed an external adapter provides.

■ *LAN Routers*

In stepping up in function, the next class of computer connections are the network adapters. These devices connect directly to your LAN or inside a computer that will act as a router or bridge on your network. These devices often have a built-in NT1 and may provide voice and S/T ports to let other devices share the circuit. When using these devices, you are usually connected to them at 10 Mbps rate internally. With that amount of bandwidth, it is possible to put multiple people onto the ISDN line with good response, just remember you are limited to the total speed of the circuit (4 users on a BRI circuit would get 32 Kbps average speed, but based on how many applications work, this speed will be much higher).

> **Bridge —** a computer network device that links two networks, sending all network information in both directions
>
> **Router —** like a bridge, except that it only sends the information to destinations you configure it to.

The LAN devices, because they have been built to support more than one user, can be purchased with multiple BRI ports, or with a PRI interface. In the space of a pizza box, you can have 23 ports for data services, giving rise to connecting that many people at 64 Kbps each. It is also possible to dedicate higher bandwidth to individual callers. Consider this example:

> You have several remote offices that need to be connected during the day, all operating at 128 Kbps to allow people to enter orders and do other sales and accounting functions. You have enough use to justify the circuit up all day long, but you are concerned about the wasted capacity at night.
>
> Internet access is something you have wanted to explore, and you have some key employees you want to offer this benefit to. You make arrangements with a local Internet provider to give you an account, and you connect your site to the Internet after hours. This access only takes one BRI circuit. You let these key employees dial into the BRI lines that are free, either through analog modems or ISDN adapters, and then they will be routed through your network to the Internet, at no extra cost to you.

■ Network Hubs

At the highest end of this product line, you have devices that act as communications gateways, where traffic from your voice, data, and video applications are consolidated into a private network backbone that uses broadband ISDN to network your sites. Studies have shown that these devices can reduce your communications costs because they more effectively integrate and transport your information than traditional lines today. You'll hear more about these exciting devices in the mixed applications chapter.

Videoconferencing Systems

Videoconferencing ranges from the complex conference room systems to the simplest of the products that are installed into a computer for a few hundred dollars. While the quality varies widely in these solutions, they all depend on two things to improve the performance. The first is the level of compression that can be done on the video signal to reduce your data transfer needs, and the second is the speed of the line, and higher is always better.

Compression capabilities have come a long way in the last few years. Systems have dropped from requiring 1.544 Mbps circuits for full motion video to 384 Kbps today. Industry prophecies say we will see 128 Kbps full-motion solutions this decade — I hope so, but it's hard to accurately predict the future.

■ *Conference Room Systems*

For stand-alone conference systems, you will typically have an ISDN interface in the video conference system controller. This device will act as an inverse multiplexor to give you access to the amount of bandwidth you need for the call. At the higher end of the market, you will have units that can be programmed to share bandwidth with telephone systems or other applications on a time-of-day basis.

If you are going to multicast your videoconference, that is, broadcast to more than one site simultaneously, you will need a

Explaining the Hardware

special Multipoint Control Unit at the primary site, which will establish connections to each remote site. Service bureaus help solve this problem by letting you distribute your broadcast through their facilities.

■ Desktop Systems

The hot topic today is desktop videoconferencing. This is about as close to the picture phone as we have been in years. Creative solutions have delivered products that require several cards to be installed. There are some solutions with only a single card, or an external device that connects to your PC and your BRI line. These solutions go up to 384 Kbps and give you full motion video if needed. Most of the desktop market has settled in at 128 Kbps, providing a reasonable range of motion and voice synchronization.

■ Picture Phones

And picture phones are finally here. Several phones have been released over the last few years that provide pretty good low-motion video. These phones are designed for people looking into the camera and moving their heads just a little, giving a great picture at the other end. Transmitting still pictures works best on the early models that could only use a single B channel, a nice way to shares ideas. As compression standards improve, the merging of desktop videoconferencing with the picture phone products will occur, bringing prices down even further.

Picture phones will be stand-alone units, just like any other ISDN phone, except they have a built-in video camera and display. The displays have been about 3.5 inches square, using LCD technology, giving a sharp picture in this small format. These units typically are not designed to share the ISDN circuit, so you will have to use some other techniques to make the most of an ISDN line connected to this device. You see some of those techniques in the applications chapters.

LCD — Liquid Crystal Display.

■ *Multimedia Conferencing*

The concept of a whiteboard within the computer environment has evolved almost to an art form. It is now possible to write on a board and have that information transmitted real-time to other sites, local and remote, and to send voice and video at the same time, to the other locations. With this technology you can perform some powerful meetings and training sessions across almost any distance. The new term for this technology is multimedia conferencing.

This equipment typically must be connected to a computer, which is then connected to your ISDN line. The computer software then manages the aggregation of the various bits of information to be transmitted and moves it out to the line. On the other end everything gets broken back out into their respective pieces and delivered to the appropriate devices at that end.

Now you can collaborate by sight, sound, computer, and written form, all through a single phone call. The speed of ISDN makes this a viable solution because the line is fast enough to support full motion videoconferencing in addition to the other information that is transmitted during sessions like this.

■ WHAT'S NEXT

Now that you have an idea about the hardware and its capabilities, plus some of the requirements to use them together, it's time to learn how to buy your lines.

CHAPTER 4
Ordering the Lines

Hang in there, you're halfway to your goal of being able to play around in the really fun part of this book, the applications themselves. Now that you have an overview of the technology and the equipment that makes this whole thing work, you're going to get down to the business of ordering your lines. Let's start with the carrier-speak and work through the issues and details for getting your lines ordered successfully.

■ LANGUAGE OF THE CARRIERS

Like every other good profession, this industry has a unique language that many of us just don't understand, and to be honest with you, most of us don't want to, or need to, understand it.

What does help is at least recognizing the terms and being able to stay up with a conversation when the people in the project switch dialects with you and start spewing out all these strange terms. At that point, you can slow them down, ask a few questions to bring it back on track, and maybe even surprise them when you understand something they said in telephonese that you were supposed to miss.

Basic Terms

There are a number of terms you are going to get in the next few pages, be patient with them, they are here as reference material to help you close the loop on things you have learned so far, and to tie together all of the literature about the phone lines and equipment you will read from the suppliers in your project.

There are a few specific terms that you want to get comfortable with; they're important to understanding how your services work with the rest of the world. While these may have been defined briefly in previous chapters, they are worth repeating here.

■ DN

The first of these is the Directory Number. This is the same as your telephone number today. In today's world each line has one number assigned to it, and additional numbers can be associated with it only through special types of service. Let's look at the following example, explaining how a single number would have multiple appearances to the world, and how it would be handled with and without ISDN.

The DN acts only as a pointer to the line the call should use, and to identify the terminal adapters and the feature keys that should be activated in the event of a call. All of this is done through a mapping process in the CO switch where your ISDN phone configurations are recognized and used.

■ SPID

Another ISDN term you have to know is SPID, or Service Profile IDentifier. In today's ISDN environment, this number is assigned by the carrier and used within their switches to recognize your device and to then work with it according to the configura-

> You have a small office and you have a few lines to handle all of your calls. You set up a Centrex system so you didn't have to buy equipment to transfer and conference your calls. You did this a few years ago, before things like Caller Line ID appeared to help you with call routing. You installed two 800 numbers, one for your local area, another for out of state. Both come in on the main line, and your receptionist answers the calls and routes them to the appropriate desk.
>
> This is getting to be a heavy load for your receptionist and you want to bypass that position with these calls. Your friendly sales rep for the telephones says he has just the ticket, a small key system with all the features you need to do this, and it's only going to cost you $8,000. Gasp — Heart Attack — WHAT! Can't afford that.
>
> Enter ISDN. With the ability to have up to eight devices at the end of the line, you could assign a unique directory number to each extension in your office. In fact, you can have up to 64 directory numbers (8 per device, 8 devices per line) per BRI line. Then when a call comes in that is directed to that extension, the call can be picked up there, without the receptionist getting involved. Your ISDN phones aren't inexpensive, this solution will still run you a few thousand dollars, but it opens the door to many other features you otherwise wouldn't have — so it's worth looking at, and paying more than you want, but less than you would with a non-ISDN solution.

tion you requested. Without this number, the central office switch will ignore your equipment. This is often a source of headaches during an installation when the wrong SPID is entered and the equipment isn't recognized by the phone network.

In situations where SPIDs are required, you will have to have a SPID for each device you need to operate on the line. For example, the Digi International Datafire card is capable of 2B data networking. Each channel must have a unique SPID. You also must have an NT1 to connect this board to your ISDN line. If that NT1 has an analog port you would like to use, you must have the phone company assign another SPID. It will also be helpful to assign another DN to allow the NT1 to properly route the incoming call to the analog port.

The physical format of the SPID varies greatly among the carrier and switches. Which format applies to your site depends on the current version of software and manufacturer of the central office switch you are serviced by. There are some common for-

■ THE ISDN CONSULTANT ■

mats that you should look for, and if you want additional information, drop by the Intel web site at http://www.intel.com/ and look at the ProShare information. Intel placed on this site a couple of pages of SPID formats for the United States and Canada. Generally, you will find that the main DN for a service on the line will act as the SPID plus some additional digits. The variations might look like this for a DN of 555-1212:

1. 01-555-1212-0 (01-DN-0)
2. 01-555-1212-011 (01-DN-011)
3. 800-555-1212-0111 (NPA-DN-0111)
4. 800-555-1212-0 (NPA-DN-0 or 1)
5. 800-555-1212-01 (NPA-DN-01 or 11)

> **NPA —**
> **Numbering Plan Area or Area Code.**

This list is much, much longer, because the schemes that can be used are based on the needs of the individual carriers. There are cases where no SPIDs are required up-front, or for operation. Instead, the next term covers the actual core of the addressing for messages, and that is the TEI.

■ TEI

The Terminal Endpoint Identifier (TEI) is a unique address assigned by the central office switch, to the individual terminal adapter at your end of the ISDN line, that will handle the call. The TEI is combined with the Service Address Point Identifier (SAPI) at the switch end, creating a unique address for your link while the call is active. Generally, the TEI is dynamically allocated at the start of each call, although it may be a permanent assignment, again at the discretion of the carrier based on their system design.

It is this combined number, TEI and SAPI, that acts as the address for all of the packets of data (voice, video, or computer data) and keeps your information flowing down the right pipe. It is the DN that gets the call started.

Voice Features

In carrying our bewilderment for the number of acronyms in the phone world forward, the names of the basic services delivered by the phone companies are often baffling. Over the decades that phones services have evolved, the industry has created a complex model for how something as simple as a call from me to you should be handled. In that model comes hundreds of possible steps and actions, where each needs to be precisely defined and uniquely identified.

Don't panic; throughout the book I am avoiding all of the details in these areas. I do so for two reasons, first to keep you focused on the exciting features that you will want to use, and second, I know that in Appendix C (the books) and on the CD-ROM (the NIUF Applications Catalog) are numerous resources that do go to that level of detail to supplement your knowledge gained here. So let's get to the exciting features in the ISDN services.

Voice features are those services deliverable on an ISDN line that go beyond the simple task of putting a call through and letting you answer it, talk, and then disconnect. These services give your phone system its creature comforts. Here are some of the best features offered today (see TABLE 4.1).

TABLE 4.1 Voice services under basic rate ISDN.

Feature Name	Description
Directory Number Appearance	Through this service you can have more than one directory number appear on your phone. This opens up the ability to have an operator's console with only one line going to it, handling 40 calls.
Call Forwarding	Automatic redirection of a call to another extension.
Call Forwarding Interface Busy	You're on the line, the call will roll over to another line, your voice mail, or your backup.
Call Forwarding Don't Answer	You're not at your desk and the phone doesn't get answered, so it will be routed to another line, your voice mail, or backup.

TABLE 4.1 Voice services under basic rate ISDN. (Continued)

Feature Name	Description
Message Waiting Indicator	If you have a voice mail system that works with the ISDN services, you can have a message indicator button that will light when any messages are in your mailbox.
Hold/Retrieve	If you need to pop between calls, speak to people in private, or set things up for a conference call, this is a key feature to make that happen.
Three-Way Conference Calling	You and two others, beating the bushes for your next great marketing idea, or discussing your losses after last week's game.
Transfer	When joined with Centrex features, anyone can route a call to another desk within your calling group, or outside the company, with the right services.
The next set of features are available through the Call Appearance Call Handling Electronic Key Telephone Service (CACH EKTS)	
Multiple Virtual Lines	Like the Multiple DNs only more powerful. These directory numbers can also appear on other phones, just as they do in a phone system arrangement.
Call Appearance Call Handling	The tools to handle more than one call at a time for a single DN, in the same fashion as an operator's station does on a phone system.
Intercom	Internal paging services, phone to phone, or individual to a group.
Bridging/DN-Bridging	This allows other phones with the DN you are calling on to know you are on a call and to join your call if permitted.
Abbreviated Ringing and Delayed Ringing	Abbreviated Ringing causes a phone or group of phones to ring immediately when a call is coming in on a line they have, but after a set time, if none of these phones answers, then the call will begin to ring on a Delayed Ringing group of phones.
Additional Features	
Calling Number Identification	Who's calling now, friend or foe? The name or number of the calling person.
Redirecting Number Delivery	Where did this call roll over from?
Automatic Callback	They hung-up, I wondered who it was? I'll let the phone dial them and see.
Six Party Conference Calling	Non-EKTS only, up to six people.

As you can see, you can tap into the power of the phone company's system, and turn their giant machine into your personal phone system. The features are growing for handling voice calls, so at the BRI line level, you can begin to see tremendous potential for sophisticated phone services with very little in the way of lines, expense, or effort on your part.

Most of the emphasis on these services has been put into the BRI circuit, because it requires the least amount of technology to deliver and use at your site. There are a number of voice features available to PRI users, but PRI is expected to terminate in a PBX or another device that will manage many of these functions. So that leaves BRI as the workhorse to bring the individual large phone system services without a system on-site.

Data Features

You certainly wouldn't expect data circuits to have a lot of features, after all, you dial a location and connect, the data goes end-to-end, and when it's all done, you disconnect. Why would you need any other services, especially any like the voice side of the house?

Many of the features listed for the voice services are offered for data circuits. In TABLE 4.2 you will see some features listed for both circuit-switched data and packet-switched data. Remember that circuit-switched calls establish a dedicated link between you and one other site, while packet-switched calls allow for multiple sites to be reached or received.

There are some fairly interesting ideas that arise from these types of features. Let's look at the next application and how a particular group of features might be used.

Again you find that ISDN has some powerful features that can support very sophisticated applications. In the data applications chapter you will find a number of examples to spur your imagination, and to solve your challenges.

TABLE 4.2 Selected data circuit features.

Service	Circuit-Switched	Packet-Switched
Calling Number Identification	✔	✔
Multiline Hunt Group	✔	✔
Basic Business Group	✔	✔
Incoming/Outgoing Calls Barred		✔
Incoming/Outgoing Call Restrictions		✔
Closed User Group Security		✔

You want to provide some telecommuting services for a few people in the office. In checking with the phone company you have found that the public packet network is connected to the central offices that will carry the traffic for each person. This opens up the ability to use this service to operate 64 Kbps packet services between each site.

One of your primary concerns is security; you don't want just anybody getting into your network, so what can you do?

By configuring the circuits with both incoming and outgoing call restrictions, you can have the phone company screen unwanted visitors from your site. And because you are using packet services, you can use only one line at each site, each configured with a single B channel for packet-data, and then use the second B channel as a voice line.

If everybody is close enough to your office, you can configure these lines into a Centrex arrangement, so they appear to be part of your office. This will allow you to transfer calls between homes and office, giving your callers the appearance that everyone is in the office. Another bonus pops out from the Centrex arrangement, your phone costs between locations is fixed, with no additional usage charges.

Capability Packages

The features discussed in the previous sections are being grouped by a *relatively simple* capability package system that has been worked out by the industry and the North American ISDN Users' Forum. More equipment is now being supplied with the capability packages identified that it will support. You need to verify that the equipment will provide the capabilities you need to fulfill. You will see capability package and order code used interchangeably by people, the official designation for this code is the former.

In TABLE 4.3 you have the capability packages, simply labeled A to T. Each defines a specific set of features and services to be provided by the telephone company. Today, you have to tell the phone company which order code you need; tomorrow, with the roll out of National ISDN-2, these order codes will be built into the equipment and the phone systems, allowing the equipment to self-configure with the phone network when you connect it to the wall.

In addition to the basic characteristics shown in TABLE 4.3, additional information is defined in the NIUF web page, http://www.niuf.nist.gov/niuf/docs/428-94.html, where the full reasoning and context of the capability packages are explained.

These capability packages are being implemented as order codes and are being phased in around the country, so you may not find them in your area yet, but they are coming soon and will aid in getting the connection right the first time.

Order Process

Once you design your application and select your equipment, it will be time to order services from your phone company. You will have a few things to do to accomplish this, but it should be fairly simple once you have gathered all of the information together and mapped out what you want.

Before you finalize your project and commit all of your funds and resources, you want to confirm that each central office switch you need to connect to has the features you want, supports the equipment you've chosen, and that you can get the same long distance capabilities if you need them.

TABLE 4.3 Order codes.

Capability Package	Circuit	Description
A	0B+D	D-channel packet data, no voice services.
B	1B	One B-channel data service, with Calling Number Identification, no voice services.
C	1B	One B-channel with alternate voice/circuit-switched data including Calling Number Identification.
D	1B+D	One voice only B-channel without any features, D-channel packet data.
E	1B+D	One voice B-channel and D-channel packet data, with non-EKTS voice features.
F	1B+D	This is the same as package E except that the CACH EKTS service is used for the voice features.
G	2B	One voice B-channel and one circuit-switched data B-channel, with non-EKTS voice features and Calling Number Identification for the data service.
H	2B	This is the same as package G except that the voice services are provided by the CACH EKTS service.
I	2B	Circuit-switch data on two B channels, no voice. Includes Calling Number Identification.
J	2B	Alternate voice/circuit-switched data on one B channel, circuit-switched data on the second B channel. Basic voice services provided plus Calling Number Identification on both channels.
K	2B	Alternate voice/circuit-switched data on one B channel, circuit-switched data on the second B channel. Voice services provided are non-EKTS and Calling Number Identification on both channels.
L	2B	Same as package K except that the voice services are provided by CACH EKTS services.
M	2B	Alternate voice/switched-circuit data on both B channels, with Calling Number Identification for both services.

TABLE 4.3 Order codes. (Continued)

Capability Package	Circuit	Description
N	2B+D	One alternate voice/circuit-switched data B channel, circuit-switched data on the second B channel, and D channel packet data services. Voice features are non-EKTS with Calling Number Identification on both services.
O	2B+D	Same as package N except at the voices services are provided by the CACH EKTS services.
P	2B+D	Alternate voice/circuit-switched data on both B channels and D channel packet data services. Non-EKTS voices services plus Calling Number Identification for voice and data.
Q	2B+D	Same as package P except that the voice services are provided by the CACH EKTS services.
R	2B	Circuit-switched data on two B channels, with Calling Number Identification. There are always two directory numbers with this package.
S	2B	Alternate voice/circuit-switched data on both B channels, including Calling Number Identification. There are always two directory numbers with this package.
T	2B+D	Basic voice services on both B channels and basic D channel packet data. This package will have three directory numbers.

When all the pieces have fallen into place, it's time to follow the order checklist to get your lines. TABLE 4.4 lists the steps that will contribute to your success in this process. Also, on the CD-ROM you will find the Order Checklist and a BRI order form in the Forms subdirectory.

TABLE 4.4 Order checklist.

✔	Step	Tasks
✔	1	Gather all of the equipment model numbers and capability package codes, list each of them by location and line they are to connect to.
✔	2	Look for any conflicts in the way you want to connect the equipment or use the line.
✔	3	For each line determine the number of directory numbers (DN) you will need for your application.
✔	4	List each line by location, with the number of DNs needed, and the capability package that provides service for all equipment on the line. Also, record the long distance carrier for each line, and any packet data services you need to reach.
✔	5	Place the order with the telephone company, providing them with this information, as well as the location of the equipment, your installation deadline.
✔	6	Record the order number given you by the phone company. Run down your questions list and make sure you have all the information you need to configure your equipment prior to the arrival of the installer.
✔	7	About two weeks before the installation due date, verify that the order is properly processed. Do this by comparing your notes with the written confirmation you received from the phone company, or following up with the people who took the order. Make sure that any discrepancies are resolved now, before they impact your installation.
✔	8	The day before the installation, confirm with the phone company that they have verified the line to your premises and it is ready for your ISDN line installation. Many of the phone companies don't do this until the day of installation, and if problems are found on the circuit, your installation is delayed while they repair it.
✔	9	Configure your equipment based on the installation instructions and information provided with your installation order. This may include programming your equipment with the DNs, the SPIDs, and possibly the TEI. Also, make sure your equipment is installed in your computers already, that the equipment responds and reports a down line.
✔	10	Confirm with the installer that your information on the DNs, SPIDs, switch type, and order packages matches his records.

TABLE 4.4 Order checklist (Continued).

✔	Step	Tasks
✔	11	Test your equipment while the installer is on site, make sure any problems are resolved then, saving you a repair call and giving the best support at the time you need it. You don't want to start with a new technician unless it is absolutely necessary.
✔	12	File away all of the information and documentation on this project. You may need it later if you have to replace equipment, do an upgrade, etc.

■ CROSSING CARRIER BOUNDARIES

When you are ordering voice services across carrier boundaries, it is as simple as picking your long distance carrier today, you specify the carrier and poof, the job is done. Even circuit-switched data will be this simple in most circumstances, but some limitations will appear and must be considered before implementing your application.

Not everyone has fully implemented SS7, so you may have to reduce your data rates to 56 Kbps to match the weakest link in your communications chain. This is a significant issue when planning networks, because the equipment and services across the entire link must be matched in order for the application to work. There may also be times when a site you need to connect to has decided to reduce their connection speed to 56 Kbps for consistency across their network, even though their equipment, carrier, and your equipment are capable of 64 Kbps.

One last implication of the reduction of speed to 56 Kbps is the fact that the network is now carrying the signaling in the same circuit as the data. This situation creates a delay in the call setup time from a fraction of a second to several seconds or longer. In a dial-up network, where the network is expecting very short connection times, this situation could force you to adjust settings and performance expectations for your application.

Another problem, serious in scope, is the ability to use the public packet networks. The use of packet-switched services depends upon the connection of your central office switch to a public packet network. If your office is connected, or can be for a reasonable sum, then you can transport your data through this effective mechanism, but many central offices are not connected to a larger network, and they can only carry traffic between locations directly connected to them.

On a final note, the issues of compatibility between the various implementations of ISDN across switch types and country boundaries are handled at the levels of the phone company switches. Within your application you only have to address the issues of communications device to device, using the protocols discussed in Chapter 2 and the line speeds offered by the carriers.

■ TARIFFS — UNDERSTANDING THE SERVICE OFFERINGS

> **Tariff** — the regulatory document covering the deliverables and prices to be charged for your services.

As you dig in and do your homework on the costs of ISDN lines, you are going to find what appears to be a number of inconsistencies among the various tariffs that dictate your costs for ISDN.

Foremost will be the vast discrepancy in rates between the various areas of the world. Several factors contribute to this situation, and while you may not be able to influence these directly, if you understand them, you can at least deal with them a little bit more knowledgeably and comfortably.

Tariff Influences

The major influence on tariffs are the regulatory agencies themselves. In many areas, there are Public Utility Commissions (PUC) that govern, in the public interest, the monopolistic conditions that the telephone carriers have operated under. Even after divestiture in the United States, and the creation of many local exchange companies, a monopolistic condition still exists for two reasons.

First, the physical wire plant to the customer's premises belongs to the local carriers, not the long distance companies or other competitors. So no matter what you do, the call eventually has to come down that wire. The old ways are almost gone as technology has provided alternative ways to carry a signal through satellites, microwave, cellular, etcetera, and the new carriers have been bold enough to begin laying down their own networks, starting the duplication of infrastructure to your site. With the next phase of deregulation in the United States it is possible for carriers to lease and eventually buy out or build access to every location, fully duplicating the existing networks.

Regulation has contributed to this monopolistic environment also. In an effort to protect the public from the perceived ravages that a monopoly can cause, the PUCs have acted cautiously to open up the markets to allow for even competition between carriers. And many carriers have used this situation to further their advantage as the regulatory mood has changed. But all of this is rapidly changing as well.

Alliances are forming globally between telephone companies in each country and counterparts overseas. Alliances, like Stentor in Canada, have brought together local operating companies into coalitions to challenge the competition from multinational companies. And the entry of cable TV companies into the networking arena signals the truest sign of competition, because they are positioned best with nearly equal access to the customer's premises today, not years ahead.

The telecommunications act of 1996 has opened the door to mergers and diversifications. Many of the U.S. regional operating companies have formed partnerships with cable companies and other organizations to prepare for entry into new markets and technologies. The recent announcement of the SBC and Pacific Telesis merger is the first of what should be a number of consolidations in the industry.

Regulatory Process

While the specifics may be different in each jurisdiction, the process of regulating activities remains fairly universal. Let's look at a brief description of the process so you see that there is an opportunity to effect this process.

1. The telephone company or PUC will draft a tariff to describe the services and goods it wishes the telephone company to deliver. Contained in the filing will be cost and benefits analysis, impact to the operating condition of the company, effect on the consumers of the product, and various details about how this will affect the marketplace in general.
2. Upon the filing with the PUC, the tariff proposal will be made public and is subject to comment and revision. It is in this time period that the public can contribute to the support or opposition of a pending tariff.
3. After receipt of public comment, the PUC will weigh the arguments presented and issue one of several directives.
    a. Deny the petition and force the carrier back to the design board.
    b. Postpone the tariff pending additional comment.
    c. Approve the tariff in its entirety.
    d. Approve the tariff subject to restriction or deletion of disapproved items.

While it may seem that the PUC may act in harmony with a telephone company, rubber stamping tariffs, this is hardly the case. Especially since the advent of the Internet and other on-line services. The ability to mobilize people to speak to an issue has increased beyond most people's wildest dreams. Through well placed electronic mail messages, large letter writing campaigns and public media exposure brings these issues to the attention of the public, and have directly contributed to tariffs being approved and denied.

A good example is the recent round of tariff proposals in California. Initially, the tariff filing established increases for both base rates and usage charges. As soon as the word hit the streets, the day the tariff was filed, cries went out on the on-line forums and the Internet. A protest site was formed and people began their lobbying efforts. In the middle of this effort, the PUC then announced that ISDN had to be treated as a resell product, forcing another look at the rate structure. As a result, Pacific Bell rethought the rate structure, and created a new rate, with an increase only to the base rate, and no change to the usage charges. While no one is every happy with a rate increase, by appearances the new tariff will be a moderate increase to the customers that were most concerned.

Tariff Components

Each tariff will contain components that you need to understand so you know what a telephone will deliver and what commitment you are being expected to make. Whether it is a restriction of the service, an unusual installation charge, or a waiver of nonrecurring charges to seed the business, all of these components need to be understood in terms of their longer-term impact.

■ Goods and Services.

Each tariff will describe, mostly in telecommunications and legal terms, the specific features that are to be covered under the tariff. Each feature description in the tariff defines the deliverable the customer will receive. This definition will encompass the action of the services, the material requirements, the customer requirements from an equipment perspective, and other items.

Translating these service offerings are often best left to your telecommunications advisor, allowing him or her to sift the menagerie of terms, translating them into common language you can understand.

What you need to watch out for is the exclusion of services you need, especially in combination with other tariffs in your area or elsewhere. You also need to understand the shift this represents from what you have today and where you want to be. It is helpful to map out the tariff in terms of your needs and see if these services fit your needs and are something worth implementing.

■ *Pricing and Commitments*

For many services today, you will find a coupling of agreement length, or term, with the price you will be charged. This relationship is built into the model of the tariff, allowing the phone companies to amortize the cost of delivering this service over a longer period of time. This is an important issue for them, because it is giving them some room to reduce your cost up-front, when they have a larger investment to make, and then allows them to make a reasonable return on that investment in an acceptable period of time.

Another factor comes into play here, continuing revenue streams. As this industry evolves to a deregulated model, the somewhat guaranteed incomes that have been enjoyed in the past go away. Brand loyalty only goes a certain distance, and since switching companies for phone services can be done in a moment's notice, some way of locking you in as a customer is necessary.

Take the pricing and then estimate your use of the services over the time period required in the commitment. Make sure that if you commit to the required term to achieve a lower monthly cost, that if you have to change your service, you are not hit with a substantial early termination fee. You will often be tempted to take the lower monthly fees, or the waived installation charges, in return for a paper commitment to two years of service, but this is not always the right way to go. Leaving early could have you paying most of those charges after all.

If you think you are going to relocate in the middle of an agreement, or that you want to move up on the technology ladder, check on your ability to do so with no penalty while still getting the best prices. You may be surprised at the flexibility of the tariffs or the company offering you the service.

Final Comment

Don't assume that you have to operate under the tariffs offered to the general public. If you are a large concern, with significant telecommunications needs, there are ways to get custom services priced and delivered to meet your needs. This is an opportunity for you to gain from specific commitments and partnering with your telecommunications company.

And with the ability to build strong private networks, you can consider deploying your resources in areas that give you favorable costs and services, and move the calls and data to that area in a private circuit. This option allows you to *look local* to your customer while you are taking advantage of business and living conditions many miles away.

■ WHAT'S NEXT

Now that the fundamentals are under your belt, we can take a look at the process of designing your application. In Chapter 5 you will learn how to determine your overall needs and then match those to a possible solution. It will be just as important for you to determine that ISDN is not right for you as it is to confirm that it is.

CHAPTER 5

Designing Applications

When I talk about designing applications, it is really the process of designing the kind of information flow you are trying to achieve. Whether that information is on paper, in a database, or just your voice, there is significance in the content, and there is a required flow and control that should be applied to it.

In order to accomplish an effective design you must define your problem, determine the existing process and capabilities of your systems, create a logical diagram and definition of what you want, identify the differences between today and where you want to go, and then create a design and plan to acquire and implement your solution. Only when all of these elements are in place will you know that every effort was made to get things done right the first time.

It's worthwhile to spend some time reviewing the process diagram in Figure 5.1. This process will help guide your activities during the entire cycle of solutions development and implementation. There are three main areas: Problem Definition, Solution Development, and Solution Implementation. Always complete this process by reviewing your solutions after they have been in operation and then determine where you need to start to continue to improve. Now let's get on with the task and learn the ways of developing solutions to our problems. In this chapter you will cover the first two areas, Problem Definition and Solution Development, Chapter 6 will go through the implementation and ongoing support areas.

■ THE BUSINESS PROBLEM DEFINED

It's great to hear about a new technology and then want to jump into the fire to use it right away. Many times you will find yourself thinking: This is just the ticket, all my problems are solved with just that one idea. But stop and take a moment to reflect on what the technology will do and then carefully plan how you will use it.

Like many great technological innovations, the ideas presented when you learn about the technology are just the tip of the iceberg, with benefits and problems hidden well beneath the beautiful and entrancing surface.

Your first step is to define the problem(s) you need solved and then to understand if ISDN will help you realize your objectives. While this sounds simple enough, often the hardest challenge is the act of defining your needs. Having clear and concise definitions of your needs is the only way to cover most of the issues that need to be in a solution (notice I didn't say all, this is reality).

Who Defines the Problem?

Many times you will find that the executive in charge will define the problem, and often that definition will fall short of the real needs of the people who have to make the solution work. One of the best ways to avoid this dilemma is to bring together

DESIGNING APPLICATIONS

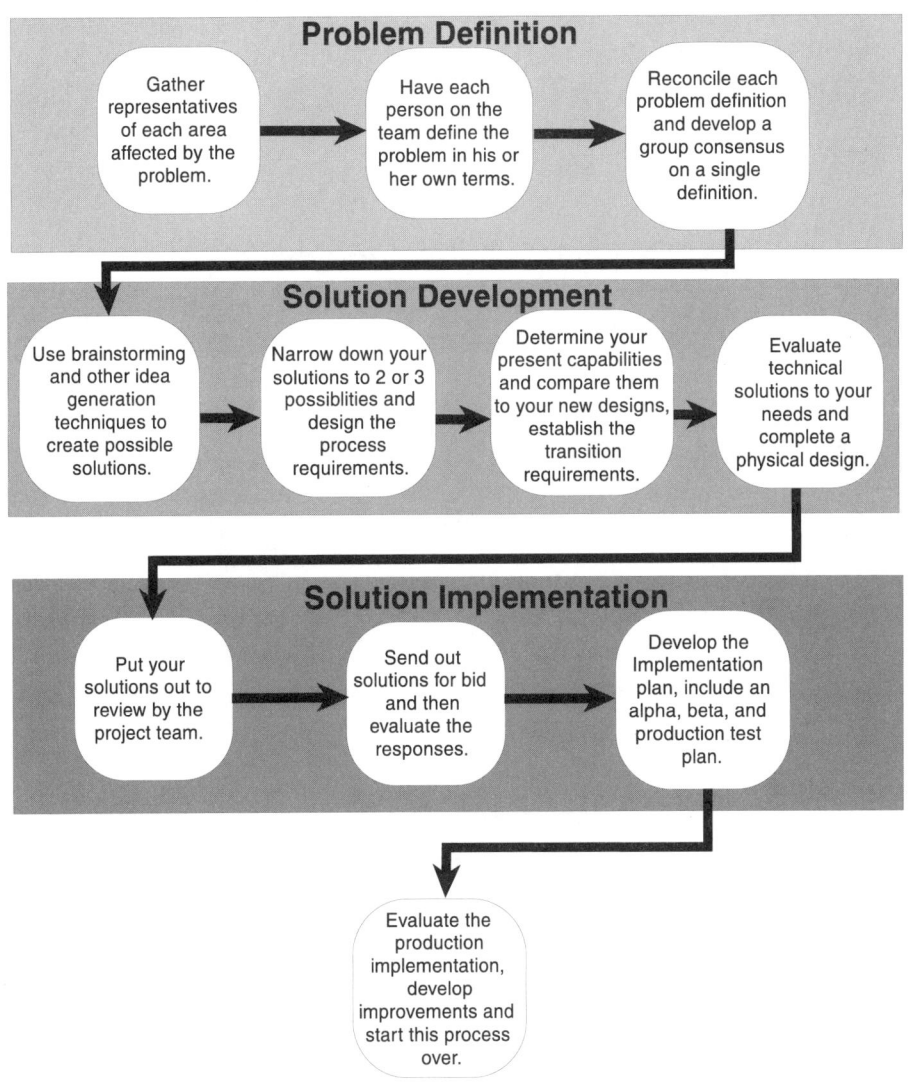

FIGURE 5.1   PROBLEM-SOLVING PROCESS DIAGRAM.

all of the people affected by the problem or responsible for the solution. In this framework, more ideas for solutions will be created, and the more likely it will be that details crucial to the success of the project will not be missed. The concept derives from work teams and has seen a lot of success when done right. For more information on this topic, look at *Exploring Requirements, Quality Before Design*, by Donald C. Gause and Gerald M. Weinberg. The authors have done a very good job of helping you understand all of the techniques and issues that go into making these activities a success.

What Does a Problem Definition Look Like?

Good question! There are a number of elements to a successful definition of a problem. Covering your bases and including as many of these as possible will help you to find the right solution to your problems. TABLE 5.1 lists each of those elements and briefly defines them. On the CD-ROM you will find a sample problem definition worksheet that you can print out and use; look in the forms directory for the problems.pdf file.

TABLE 5.1 Problem definition template.

Problem Scope	Does this problem affect only a person, a department, or cross department boundaries? Is this an internal issue, or does it affect outside parties?
Reactive or Proactive	Is this problem in response to an unpredicted situation, or is this anticipating a future need?
Responsibility	Who is responsible? Every problem belongs to somebody; who is ultimately responsible for this one?
Authority	Who has authority to approve the solution? This may not be the same person who is responsible for the problem. This element helps you identify the proper person to sign off and fund your solution.
Deadlines	Every problem has a deadline and often the deadline will force compromises in the solution, so it must be taken into consideration.
Budget	What funds are available to support the solution. Always the primary constraining factor, you may have to shift focus and resources based on this constraint.

Answering these questions will take some time and effort, but they make your solution development a lot easier, and they will help to bring you buy-in when it comes time to get final approval. Let's take a quick look at a problem definition for a small manufacturing company (see TABLE 5.2).

TABLE 5.2 Manufacturing problem definition.

Problem Scope	Two things have happened to create a need for high-speed communications. The first is the influx of new business, where there are large engineering drawings to be transferred from the customer's site, modified here, and then forwarded to a supplier for additional work. The second is the desire to create a remote office closer to home to work on engineering drawings, saving a 50 mile commute one day per week.
	The new business is to begin in three weeks, and will require transferring up to 5 MBs of files each day. These files will be modified, combined with other drawings and then sent to the supplier each day. Total transfer needs are for 10 MBs daily.
	Any changes to the communications technology will require compatibility with the customer and supplier.
	For telecommuting, the need is to transfer the engineering drawings as required to perform the work offsite, and run any of the production systems to monitor the business, process orders, E-mail, etc.
	The telecommuting also raises the issue of desktop videoconferencing to keep in touch with the employees, or to monitor the shop after hours through the security monitors.
Reactive or Proactive	This is mostly a proactive project, except for the tight time line for the incoming engineering drawings. The existing modems can do the job, but it will take up to an hour per day to get the job done.
Responsibility	The engineering department is responsible to get the drawings done daily.
Authority	The production manager has the authority to purchase a solution.
Deadlines	FIve weeks from now.
Budget	Up to one year of cost savings can be spent on this project. If the project saves half the time it currently takes, the budget will be $6,500, if the savings is 75 percent, then the budget can be $9,750. This does not include the expense of the videoconferencing, which is seen as an added benefit and separate project.

As you can see, this isn't a really complicated problem, and the process definition is fairly simple. What you should notice is that the problem scope covered several issues, and identified a new issue outside the scope of the original problem, but within the need to include it in any design considerations. This will be typical of many projects when you are looking for the best possible path to follow.

The next step would be to take this form and have the various responsible parties review the problem and return their own versions of the problem statement. You may not gain anything new from this step; however, that would be the exception. Additional issues usually surface at this stage, and it is easier to deal with them now rather than later.

You complete your problem definition by producing a composite of the input of all of the project team members, which has their concurrence of the team and the approval of the management.

■ SOLUTION DEVELOPMENT

There are four different areas to address in developing your solutions. The first is lots of fun — it's where you generate a lot of ideas. In the second phase, you screen those ideas down to a few possibilities that you will invest your research time and energy in, saving your resources only for those ideas that appear to be most comprehensive and reasonable. In the third phase, you must compare your own reality with your proposed solutions, and then narrow the solution to a reasonable time frame and change. You finish this effort by developing the complete technical solution and agreeing on which alternatives will be sent out to bid.

Idea Generation

The purpose of this step of the process is to open your mind to new ways of thinking about a problem. There are many techniques that you can use to do this, but they are all headed in the same direction — to find a better way to do things, and to do it within the scope of your objectives.

Throughout this process, you are headed toward the position of having choices to make, instead of having only one solution. There is rarely a situation where only one choice exists, and when people are given the opportunity to choose a direction, they are more likely to support that decision. And with many technology projects, it is often the lack of feeling that there was a choice that leads to the failure of the troops to support the project, and that causes the project to fail.

Narrow the Solutions

Regardless of the techniques you choose for generating ideas, take time to record all of them, and then sift through them to find the jewels. Two things should happen at this point. The first is the rejection of those ideas that are: too far out to be considered or outside the scope of the problem, the authority, the budget, or any other aspect of the problem definition. The second is organizing the remaining solutions for evaluation.

Once you have a set of acceptable solutions, you want to reduce them to only two or three. Ranking is generally a good way to do this. Create a criteria (be certain you have chosen the right criteria) by which you can judge the merits of each solution, and then rank each solution within that criteria. When you are done, you can take the top ranking solutions and move forward.

One step of moving forward is the creation of a logical process flow diagram, and a tentative design diagram for each proposed solution. This step will give you a more consistent framework under which you can compare solutions to each other and consider them by today's needs and tomorrow's expectations. In the next two figures, Figure 5.2 and Figure 5.3, there are the logical process design diagrams for the manufacturing problem described in TABLE 5.2. Just putting this down on paper helps you out, while also creating a document your vendors can use to develop a better bid for you.

As you can see, these diagrams are very simple. Your diagrams might be this simple if the application allows, but more often you will want to include all the process steps and components you expect to use. In this situation, I assumed several things in the physical diagram that I want as part of the solution. The first

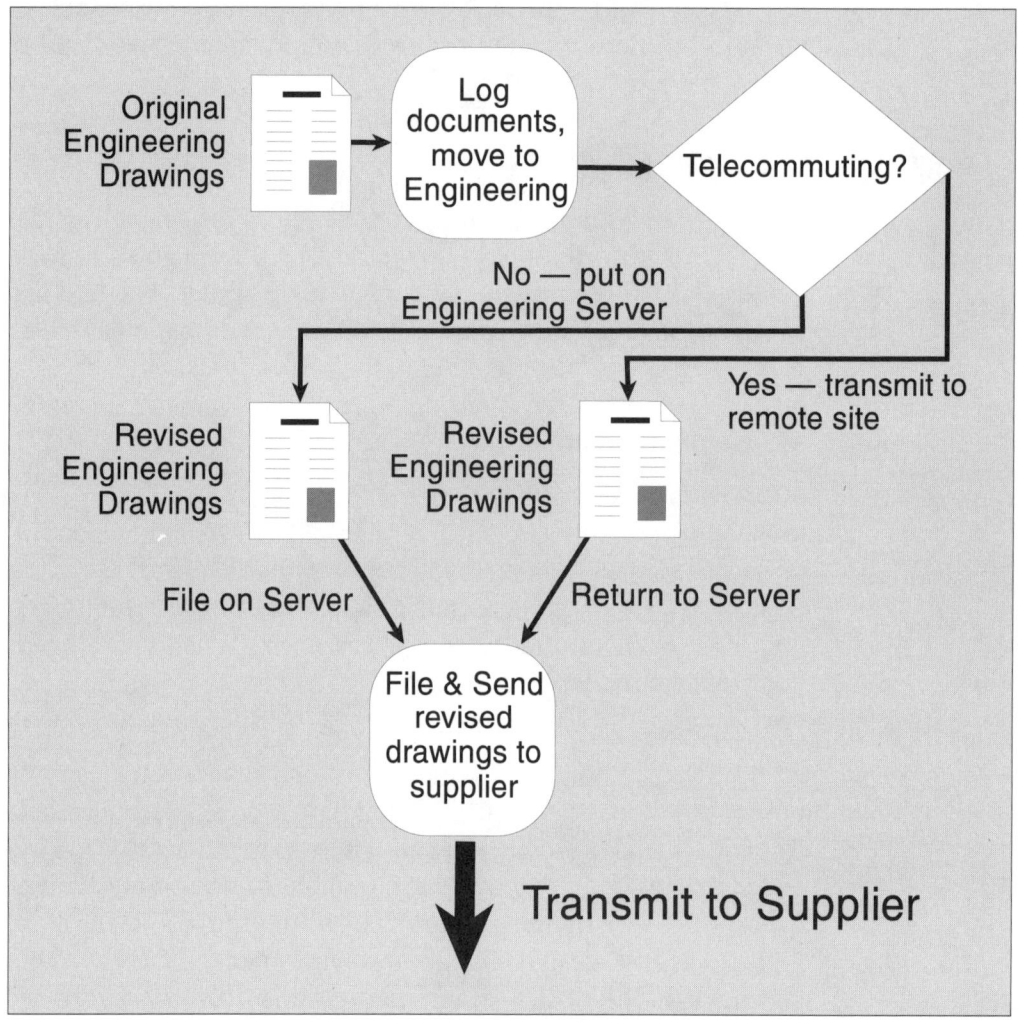

FIGURE 5.2 MANUFACTURING COMPANY PROCESS FLOW DIAGRAM.

is that the remote computers would all use internal TAs and all of the TAs, LAN and remote, would have integrated NT1s.

It can be quite easy to develop these diagrams with the right tools. There are a number of tools available to help you with this process. For this book I used netViz 2.5 from Quyen Systems, Inc. This is a network diagramming tool that also handles flowcharting tasks. With it I am able to create a picture of the process I need to

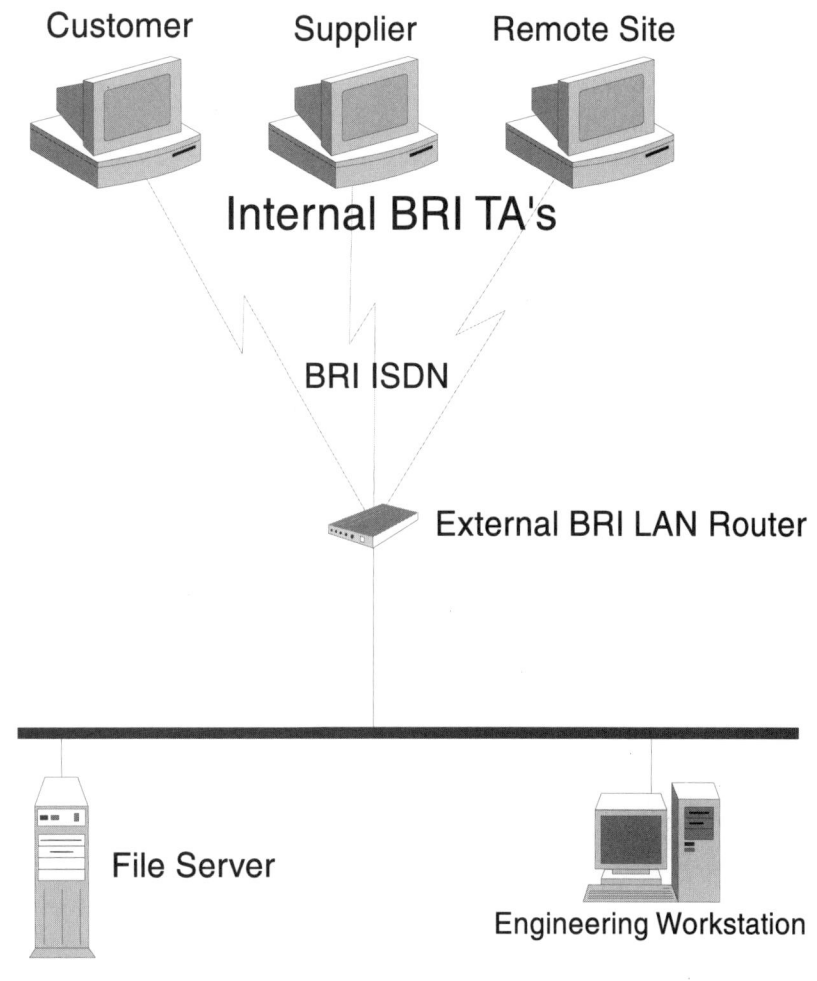

FIGURE 5.3 MANUFACTURING APPLICATION PHYSICAL DESIGN DIAGRAM.

develop a solution for, and then I can create the network diagrams for that process and link them back to the process flow. And when it is time to purchase the equipment, I can add information to the diagrams, such as the model, capacity, and other information about the network. This information becomes invaluable later as you need to troubleshoot or upgrade your solutions. Your diagram can contain all the information that is needed to resolve many networking issues.

■ THE ISDN CONSULTANT ■

To give you an idea how simple the interface to a program like this is, I have included a sample screen in Figure 5.4 and a demonstration version of this software on the CD-ROM.

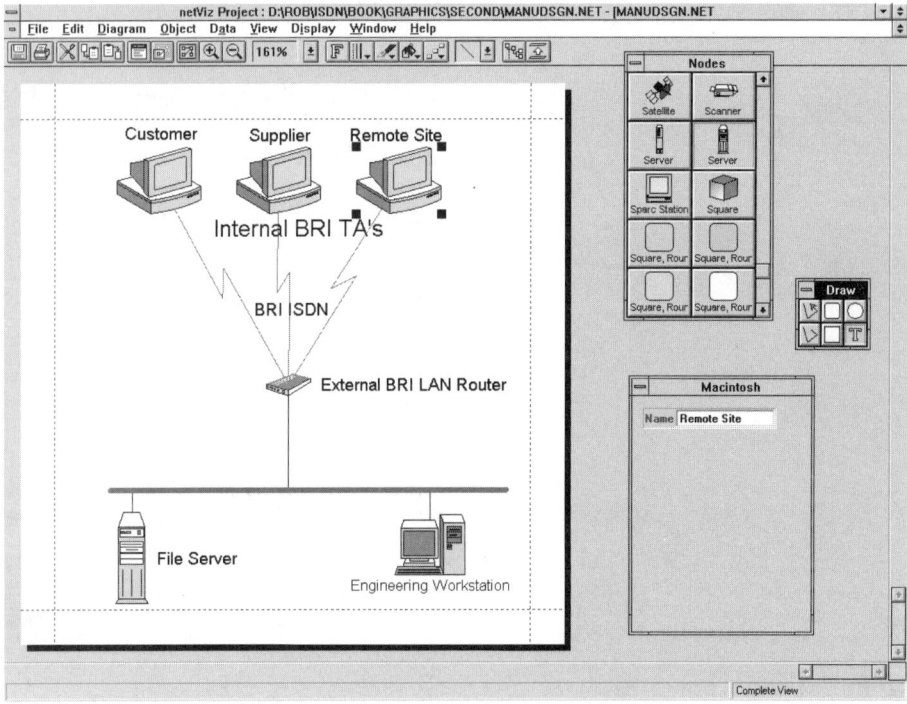

FIGURE 5.4 NETVIZ SAMPLE WORK SCREEN.

Determining the Capabilities of Your Existing Systems

Every solution is built on getting from where you are today to where you want to be. Even if you are willing to throw out the old and bring in a complete replacement, you have to understand where you are today to do it right.

Wiring

One of the major constraints in every telecommunications or computer networking project is the existing wiring. Establishing even where the wire exists, verifying the integrity of the wire, and its ability to support your future applications can be a tough challenge. Many buildings and homes have been built with poorly or undocumented wiring infrastructures. The first major expense may be determining the present state of things, or investing in a complete gutting of your wire plant and then installing something known and well documented.

Existing Equipment

Another area of concern is the capacity of your existing phone and data equipment to work with ISDN equipment. You need to realize that many phone systems sold in past years have been digital systems, but digital with proprietary technology, and that means the equipment may have no ability to add ISDN, forcing you to buy additional equipment to act as an interface to the ISDN services. This equipment purchase may make sense, but only when you understand why and how it will meet your needs.

To gather information about your existing equipment, go to the equipment manuals (I put this first because often the answers are there). Talk with the technical support people on your staff and with the vendor. Many times the person who fixes your equipment will be an excellent resource for questions like:

- Will this equipment support ISDN?
- How successful has ISDN been on this system?
- How hard has it been to install ISDN in this area and what have the problems been?

Many of the telephone companies, and even the a number of the equipment vendors, now have an ISDN help line to give you advice on what options you have with your equipment. In Appen-

dix C and on the CD-ROM (NIUF Application Catalog) there are hundreds of resources to help you out.

■ *People Skills*

Take a close look at the skill of your staff and the vendors who support you. If your staff seems knowledgeable about the subject, great — but verify their knowledge by bouncing the plans off a disinterested third party who is qualified to review this material. There will be times when you find someone who talks the subject really well, but has very little depth to back it up, and you need to screen these people out of your project as early as you can.

What everyone in your project should be able to do is articulate their part of the project, and put into perspective how any given solution or technology meets your needs. This requires that they understand the process, the business requirements, and the technologies involved. If they do, then your questions should pose no challenge for them, they may not have the answer right away, but they will get one and it will stand the test of inquiry.

Some key skills to look for when you embark on a project like this:

- Good listener — able to ask more questions rather than give immediate answers. They can communicate what you said back to you in such a way that you are comfortable they truly understood you.

- Good written communications skills — able to draft project memos, plans, and other written materials.

- Process thinker — can define your needs in terms of a process, and then link that process to the technology.

- Grace under fire — technology projects eventually get tense, and you need someone who will remain calm under fire, who will focus on the solution and root cause, not the symptoms.

- Technology literate — understands more than just a specific category of technology. A broad picture and familiarity with other technologies is important.

If these skills don't exist on your staff, then look for them in the vendor or consultant you bring in to help you with the project. If you bring someone in, use them to train at least one person on your staff so you don't lose that valuable expertise at the end of the project.

Identifying the Missing Links

Once you have completed your assessment, it's time to look at your design and evaluate how much of your infrastructure will work with this new technology. This will be the time to define the specific requirements of your solution.

There are a number of areas that can trip you when it comes to implementing new technologies. TABLE 5.3 is a short list that you should check before you make a lot of plans. It will usually contain something in an area that will present a problem. You'll notice that not all of the issues are technical, and perhaps more often than not, the biggest hurdles will be the change issues within the organization and the people.

TABLE 5.3 Issues to pay extra attention to.

✔	Issues
✔	The existing telephone system has no expansion space left. This will often be a problem that requires the addition of new shelves in a PBX, an expansion unit in a key system, or a forklift upgrade (a whole new box).
✔	The telephone system won't handle the new digital technology trunk cards. Many phone systems implemented digital technologies before ISDN was standardized. This will be true of phone systems from the early 1990s or before.
✔	The systems software must be capable of networking.
✔	The computer system you want to install ISDN into won't have enough expansion space.
✔	If you are installing into an Intel based system, your available interface settings may not work with your ISDN adapter.
✔	There may not be a software driver for your device in your operating system.

TABLE 5.3 Issues to pay extra attention to. (Continued)

✔	Issues
✔	The wiring in your home lacks enough pairs to run the ISDN circuit where you want it installed.
✔	The LAN ISDN device you want to install doesn't support your computer network protocol.
✔	Your Internet provider doesn't support the protocol that your ISDN adapter needs to use to connect to them.
✔	Your insurance company and lawyer tell you there's not enough coverage on your liability and workman's compensation policies to cover employees working from home.
✔	The terminal adapter won't supply the proper analog phone services to work with your phones or phone system.
✔	Your employees don't like the idea of tying up their home lines with a business line, or knowing you expect them to work after hours.
✔	Employees may distrust the effort to automate some functions because it seems to be threatening their jobs.
✔	Your cutover date to the new lines is too close to a significant company activity. Murphy's Law says that this will always happen when you don't check for it.

While some of these issues may seem trivial, it is often the trivial issues that will conquer your project, just as David defeated Goliath. I hope the list stimulates other issues for you to look at and helps you find the one that would have otherwise killed the project at just the wrong time.

You also need to utilize a checklist approach to each element of your current environment and identify where you plan to add ISDN circuits. Then walk through those devices' abilities to use ISDN equipment. Any discrepancies found at this point need to be accounted for in the next section, the physical design.

Determine Technical Solutions and Physical Design

If we revisit the manufacturing example earlier in the chapter, you saw in Figure 5.3 that a number of elements were called for in the design. In particular, the remote computers are Macintosh computers, placing certain limitations on the types of adapters that can be used inside the computers and how external adapters would operate. The terminal adapter attached to the network will require that the remote computers and the LAN operate a routable protocol, like TCP/IP; this will be another issue that the physical design must solve.

These are the start of questions that must be resolved as you build the physical design. During this phase you determine the technologies you can use, you prequalify the vendors that can meet your needs, and you set the framework of the final solution for evaluation by the project team. The following questions need to be answered for each device that will interface an ISDN line to your equipment. These questions will go beyond the hardware, and will also address the software components necessary to make ISDN work (see TABLE 5.4).

TABLE 5.4 Design question list.

Questions
For BRI Lines
1. How many voice Channels are required?
2. What voice services need to be available? Will this installation include Centrex services?
3. Will there be other ISDN TA's installed on this circuit?
4. If there are to be packet-switched data, are the central office switches connected to the appropriate public-packet network?
5. What are the Central Office (CO) switches that the equipment will connect to? What requirements does this impose on the connecting equipment, like 56 vs. 64Kbps circuits?
6. What type of ISDN software is the CO switch running?
7. What are the lead times for ISDN installations at each location?
8. How many analog ports are required for the TAs? Will those TAs provide the required signals, including ring, to the analog equipment?
9. For TAs that are external to computer equipment, are their speeds limited by the serial port connections?

TABLE 5.4 Design question list. (Continued)

Questions
10. If the TA is a multifunction device, with fax, phone, modem, and data capabilities, are there any special configuration requirements for this equipment to work with your CO and other equipment?
11. If you are installing multiple ISDN TAs on a single circuit, do you have an NT1 that will support this configuration, and will the CO support it?
12. Does this TA need to talk to a computer telephony interface?
For PRI Lines
1. Are there any expansion slots in the PBX to install the PRI line cards?
2. If this is an additional PRI circuit, can the existing D channel handle the card, and is there a backup D channel capability?
3. What type of ISDN software is the CO switch running?
4. Does the PBX have the capacity to deliver BRI services out the port side of the switch?
5. If packet-switched data is to be sent down the PRI, is the CO switch connected to the appropriate public-packet network?
6. Is an inverse multiplexor required to distribute portions of the PRI circuit prior to using it on the PBX?
7. What interoperability is required with computer and videoconferencing equipment?
8. Can the equipment or CO switch provide multirate ISDN, if required?

After all of the questions have been answered, you can create a detailed design document from which your vendors can bid products and services. This kind of homework allows these people to give you accurate and effective responses that you can use to properly budget your project. And if you are not up to putting all of these answers together, then it makes a lot of sense for you to hire a consultant or a vendor to do this for you. For you the key will be in knowing that these details have been taken care.

For a final look at what might be produced in the way of a design diagram after these questions have been answered, let's look at Figure 5.5, where the manufacturing application is completed for release to the vendors for a quotation.

DESIGNING APPLICATIONS

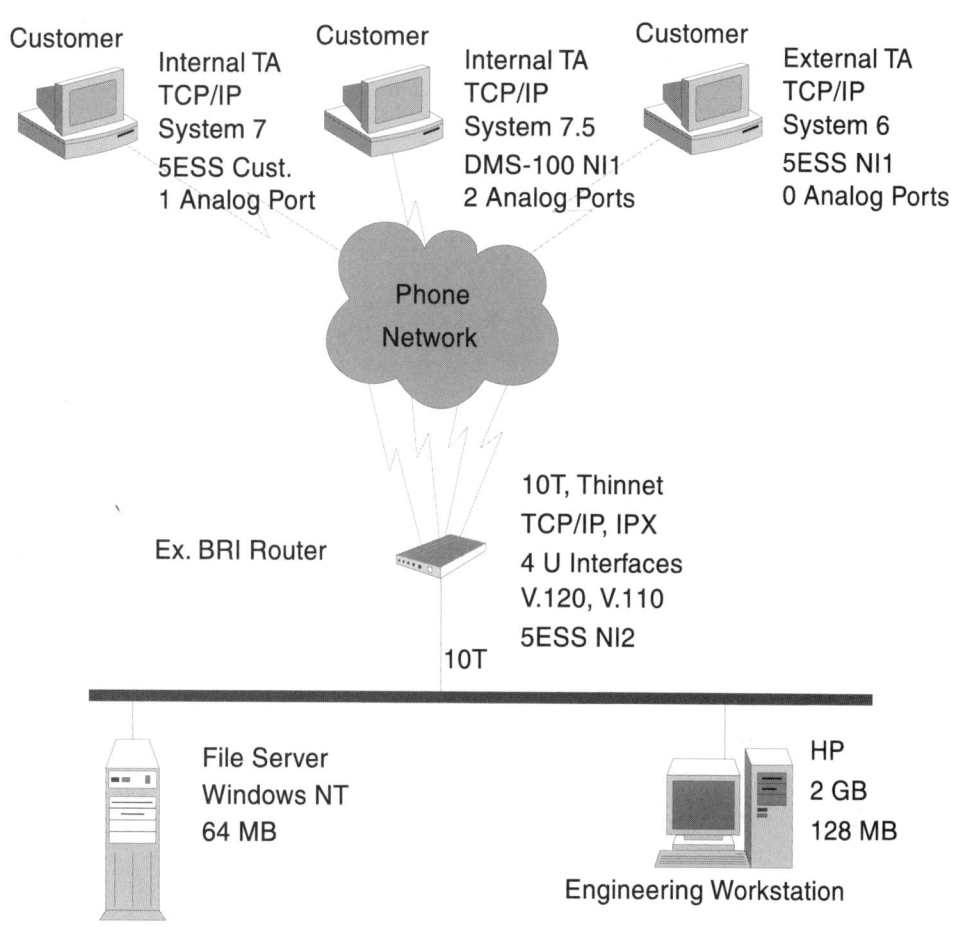

FIGURE 5.5 FINAL DESIGN DOCUMENT FOR MANUFACTURING APPLICATION, WITHOUT VIDEOCONFERENCING.

You notice that key questions have been answered and placed on the diagram. Space permitting, you should detail as many of the hardware, switch, and operating system facts as you can on your diagrams. If necessary, create several versions, to save time later in the process. Vendors will ask for that information in the context of what your network is going to look like.

 This is where the strengths of the newer network design packages come in. With the ability to collect and report all of the information about your project design, you no longer have to keep the information on scratch paper, a separate database, or elsewhere. And these databases can usually be reported on by database software, making the data that much more valuable.

■ What's Next

Can it be, the end is near? I see the light, we're getting ready to buy everything and put it in. On to the next chapter where you will finish the lessons and close the circle, going from an idea or notion you wanted to put in an ISDN solution to being fully equipped to do the real thing. After Chapter 6, you get to relax and enjoy a series of real and possible applications, all intended to help you find your perfect fit for ISDN.

CHAPTER 6

Implementing the Solution

All right, enough with all of the background material, I know what I want, and it's time to get at it. Can't I just order the equipment and install it, everything is going to work right out of the box, as easy as counting to 1-2-3, right? I'd sure like to tell you that it will, but this is a book about reality and possibility, not fantasy. Any project dealing with technology needs adequate planning to be successful, and even more so when it is a newer, leading edge technology.

In this chapter you will learn: (1) how to put your final design into a request for quotation (RFQ); (2) how to evaluate the responses; (3) to understand the lifecycle of a project and the general components of creating an implementation plan; (4) what it takes to install ISDN services (all the nasty details); (5) what to look out for, what type of testing you should do; and finally (6), how to know when the service is operating properly and what to do when it has failed.

■ FINAL REVIEW, GOING TO BID

As you go to final review with your project team, you want to put a document together that includes enough information for the reader to understand the following things listed in TABLE 6.1. This is a complete outline to guide you through the construction of a review document, and forms the basis of the bid document. There are also the sections that should be included in your bid document. I have included a brief explanation of the content of each element in this review document and a check mark if that section should be in the review (R), bid (B), or both documents.

TABLE 6.1 Review and bid outline.

R	B	Section	Purpose
✔	✔	0.0 Executive Overview	A few pages explaining the entire project and conclusions in short, precise, and clear terms.
✔	✔	1.0 Background	Explains in general why you are doing this project and the steps that got you to this point of investigation. This section should ground the reader in your need to do something.
	✔	1.1 Company Background	An introduction to your company, the business you are in, and what the objectives are in using the technology.
✔	✔	1.2 Document Format	Explain the general format of your document, any unique elements, and define your acronyms and other abbreviations or conventions used to simplify your material.
✔	✔	1.3 Confidentiality	Define the expected level of confidentiality for this document. If nondisclosure agreements from potential suppliers are required, include the forms here.
✔	✔	2.0 Problem Statement	What is the problem? This should include each 2.x section, starting with a general description.
✔	✔	2.1 Groups Impacted	Include all of the people and organizations affected by your problem and solution. This includes internal departments, customers, vendors, and employees outside normal work hours.
✔	✔	2.2 Process Changes	What process flows will be affected by this change? Does this introduce additional steps or complexity, or does it simplify your efforts?

TABLE 6.1 Review and bid outline. (Continued)

R	B	Section	Purpose
✔	✔	2.3 Technology Impact	Explain the changes to technology you require to accomplish your desired process.
✔	✔	2.4 Financial Impact	What is the expected financial benefit and cost of this change?
✔		2.5 Internal Comment	Include the feedback from your project team and other interested parties on the problem statement and associated issues.
✔	✔	3.0 Current Situation	Open with an overview of your present operating environment and how you are solving the problem today.
✔	✔	3.1 Process Flow	Provide a diagram and explanation of the existing process.
✔	✔	3.2 Technology Diagram	Show how your current technology supports the existing process. Also include information on what expansion or upgrade capabilities your equipment has.
✔	✔	3.3 Process Measures	Explain how you measure the productivity and success of the current process. Discuss if this is going to change with the new solution, or what baseline you will establish to measure the success of the new solution.
✔		3.4 Internal Comment	Include the feedback from your project team and other interested parties on the current situation and associated issues.
✔	✔	4.0 Solution Objectives	Describe why this change to your process and technology achieves the results you want. Explain why this approach will solve your problem best.
✔	✔	4.1 Solution Requirements	Here you will get into the details of your plan. Provide a process flow diagram, an expected technology diagram, and detailed characteristics about how your solution should operate to meet your needs.
✔	✔	4.2 Project Timeline	Create a general project schedule, including milestones, and a definition of firm and relative dates in the project. A firm date would be a deadline that must be met, while a relative date would be something like, this event must occur within ten days of the receipt of the bid.
✔	✔	4.3 Project Team	Who will participate in the project and what are their roles and responsibilities? Clearly define here the chain of authority to resolve issues within the project and the extent of authority the project team has.

TABLE 6.1 Review and bid outline. (Continued)

R	B	Section	Purpose
✔	✔	4.4 Supplier Responsibilities	Explain what you expect from the potential suppliers in this project. This should include all of the project duties like design, training, implementation, ongoing support, etc.
✔		4.5 Internal Comment	Include the feedback from your project team and other interested parties on the solution statement and associated issues.
✔	✔	5.0 Response Document	Define the required format and elements in the supplier response. Use this section to create a structure that will help you evaluate the responses consistently.
✔		5.1 Internal Comment	Include the feedback from your project team and other interested parties on the response document and associated issues.
✔		5.2 Response Evaluation	Identify the criteria and weights you are going to assign to each response from the supplier. Create a matrix to record the results of each response. Doing this activity helps you to understand the importance you place on different components of the solution, and will make the bid evaluation process go much faster.
✔	✔	6.0 Other Elements	This is where you provide information on contracts, legal requirements, your purchasing process, commitments you will make, expectations you have of the supplier, etc.

As you see, the elements of a design and bid document are extensive. You need to tailor this document to meet the requirements of your organization. If you are a smaller company that runs lean and mean, where the person with the authority, responsibility, and interest in this project is the same person, then skip through many of the elements, because you would be writing them for yourself. But going through this process will ensure that you have looked into your objectives and done a good job of acquiring a solution to meet both your immediate and longer-term needs.

■ PROJECT LIFECYCLE

As you plan your implementation, in which you have just invested many hours, it helps to understand the phases of a

IMPLEMENTING THE SOLUTION

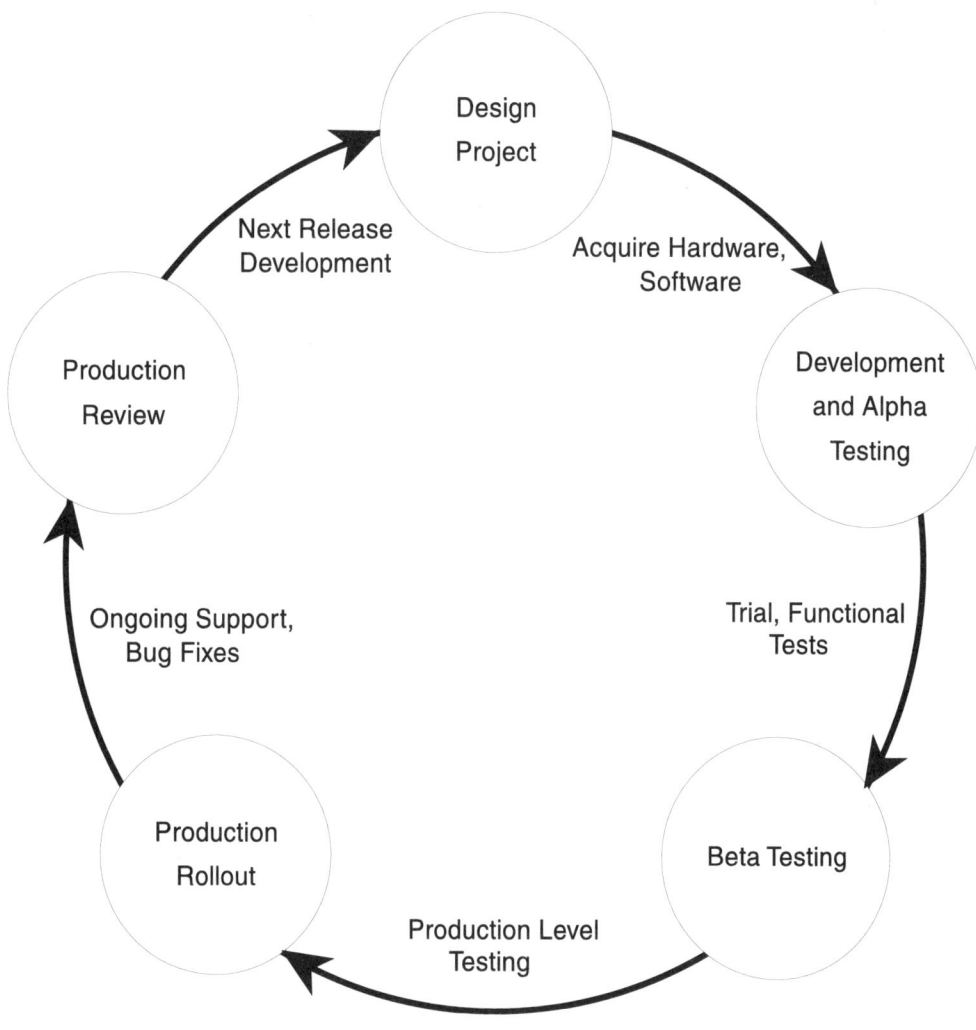

FIGURE 6.1 PROJECT LIFECYCLE.

project. It takes many days to weeks to get through the first phase of the process. In Figure 6.1 you see that a project will go through five distinct phases. Each of these phases serves a unique purpose to aid you in dealing with all of the issues that arise at each step of designing, acquiring, and implementing any change in your organization.

One of the great things about word processors today is their ability to make constructing a document like this a simple task. By creating a style, you can specify what each part of your document should look like. And in doing that, you give each paragraph type a name; which now gives you the to ability to manipulate entire sections of text with just a few keystrokes. In this situation, you can turn the entire style into hidden text, allowing you to make entire bid sections disappear, or vice versa. You could review only those areas you're responsible for, then cause them to disappear from the bid document. In the mean time, you only have to maintain one document, where you can keep control over the entire process.

On the CD-ROM is a Microsoft Word 7.0 document that is a template for this review/bid document. It contains styles for each of the areas as I have suggested them, and it has a few macros in it that will hide and unhide the text sections in the document you want to print. Feel free to modify this template any way you need to; it's just a starting point to help you create these documents in the first place.

In TABLE 6.2 a number of terms are defined to aid you in talking the language of the systems designers and consultants in the industry. These terms describe major events in each of the phases of the project.

TABLE 6.2 Project terms.

Phase	Term	Definition
Design	Process Flow	A visual and textual definition of the process you are designing.
	Technology Design	A complete visual and written representation of the solution you want to implement.
Development	Alpha Test	At this stage of testing, you test the procedures, hardware, and software with the expectation that much of it may not work the way you expect, and that it is subject to great change.
	Frozen Specifications	The initial design plan will be subject to changes during the development, because you discover issues and needs that can't be found except through this process. At some point, you need to freeze the specifications and make changes only when required to fix a bug or meet a significant change in the project requirements. If you don't do this, then you may never get past this phase of the project. There are other phases to deal with new requirements.

IMPLEMENTING THE SOLUTION

TABLE 6.2 Project terms. (Continued)

Phase	Term	Definition
Beta	Beta Test Plan	This is a comprehensive test plan that will stress all functionality of your design. It includes functional tests of the process, hardware, and the software.
	Beta Test	This is usually a parallel use of the new systems with the old. During this test phase you determine that the new systems will work according to your designs and according to the user's perceptions of improvement over the old systems. Minor changes to system design can usually be accommodated at this point, but changes should be reserved for major bug fixes only.
	Beta Review	This is a formal evaluation of the beta test. You judge the success of the new systems against your predefined performance measures; establish those last items to be corrected or enhanced to go to the release step; and determine if a second beta test or production rollout will occur.
Production	Rollout	This should be the staged implementation of your new system. Often, if possible, you want to stage the deployment of a new process or technology, to give it a little more trial by fire without effecting everyone in the organization. This is not always possible, but it is a conservative approach for implementing new technologies to minimize the risks.
Production Review	30–60–90 Day Review	A review of the success of the new system should be conducted at these points of the rollouts. In general, just like a new product that might fail in the early days of its life, a new system will tend to do the same things. These reviews help you understand the issues that will make up the next project for this process. The information gathered here, plus additional reviews later on, will give you the basis for future enhancements to the solution.

The Implementation Plan.

I remember being taught to always plan your work and then work your plan. Later in life I was told that you should be flexible because no plan is perfect, and you need to make your plan flexible enough to deal with unknown circumstances. This was sage advice then, and every time I forget to follow it, I find myself in more trouble than I would have been with a flexible plan.

In TABLE 6.3 is a sample ISDN project plan. There is sufficient detail and flexibility to use this template on any of the projects contained in this book. Follow along with the plan as you read through a brief explanation of the process and extra things to look out for.

TABLE 6.3 Sample implementation plan.

Phase	Action	Week									
		1	...	8	9	10	11	12	13	14	15	16	17
Definition	Assemble Project Team	✔											
	Define Problem	✔	✔										
	Finalize Project Objective		✔	✔									
	Determine Current Capabilities	✔	✔	✔									
	Evaluate Technologies	✔	✔	✔									
	Finalize New Process	✔	✔	✔									
	Project Approval		✔	✔									
Development	Assemble Development team		✔	✔									
	Purchase Products				✔	✔							
	Order Lines				✔	✔	✔	✔					
	Develop Alpha and Installation Test Plans					✔	✔	✔					
	Test Lines & Equipment						✔	✔					
	Develop Training					✔	✔						
	Train Alpha Site						✔	✔					
	Alpha Test 1 Site							✔	✔	✔			
	Alpha Review									✔			
Beta	Beta Plan Development							✔	✔				
	Train Support Staff						✔	✔	✔				
	Train Beta Site								✔	✔			
	Beta Test 1 or More Sites										✔	✔	✔
	Beta Test Review											✔	

IMPLEMENTING THE SOLUTION

TABLE 6.3 Sample implementation plan. (Continued)

Phase	Action	Week										
		1	...	8	9	10	11	12	13	14	15	16	17
Production	Install Lines & Equipment										✔	✔	
	Train Users												✔
	Put into Production												✔
	30 Day Review												▶
	60 Day Review												▶
	90 Day Review												▶

With each phase comes a series of project meetings to review the status of the project and evaluate its success at that point. Each review meeting held at the end of a phase is considered a milestone, where a decision is made to proceed to the next phase or delay the project pending changes and additional development. Milestones are the important checkpoints in your project, because they are the time of reflection and soul searching to know if the project is meeting its objectives and going to give you the results you want. Feel comfortable with the idea that you could delay or cancel a project based on the findings during one of these meetings.

The timelines shown in this plan are relative, you will need to adjust them according to the size of your project. For example, in a project affecting 2,000 people, with multiple systems, sites, areas, and objectives, you want to plan long lead times, consider breaking the project into subprojects to make it more manageable, measurable, and achievable. Now take a look at an example of a telecommuting program that has been implemented by a high tech manufacturing firm to enhance its work environment to attract and keep talented employees.

Keep in mind, the whole purpose behind this process is to get you comfortable with the technology, to prove out your solutions before you become dependent on them, to clear out all the kinks, and to gain acceptance of the new capabilities you are gaining access to.

■ THE ISDN CONSULTANT ■

At a high tech manufacturing firm a growing need was identified to support employees working from home during their off hours or on a permanent basis. With previous programs having used analog technologies with limited or no success (some employees gave up on the idea of LAN to LAN activities in the analog environment), the company decided to put a trial together using ISDN, based on its high speeds and general availability in many of the areas where they had operations.

To move forward with this project, an initial alpha test was designed with about 50 volunteers. The services would be made available without any guarantee of a stable environment. This alpha was to be a period of experimentation and learning. While every effort was made to choose equipment, software, and procedures to make this phase run smoothly, the task at hand was new for many of the people and, as such, the learning curve dictated that problems would arise.

With this company, tracking the productivity and the costs are very important. And each group using these services needed to be billed for their use of these shared IS resources. On top of that, there is a software management program to ensure that all desktops in the company have fully licensed software, networking configurations, and proper access to company resources, so ISDN had to be integrated into this environment.

With that challenge understood, the alpha was undertaken, evaluating hardware, software, ordering and support procedures. The alpha ran for four months. All of the issues related to installation, training, and support were evaluated during this phase. The initial models were adjusted. For example, the idea of installing all of the systems and ISDN TAs using technicians was changed to a user installation with preconfigured units and better documentation.

After a successful alpha, a production pilot was undertaken with 150 people, testing everything developed for a full production rollout. Final adjustments were made to the materials and procedures, and the program was released to the company, in that area, beginning January 1996. The program also was initiated in Europe and across the United States, where a similar process was occurring, built on the lessons learned in the alpha and beta phases of this project.

■ TESTING PROCEDURES — TRYING TO GET IT RIGHT THE FIRST TIME

Core to every successful implementation I know of is the testing of the plan and its components. The U.S. space program is an excellent example of planning and preventative measures, as a success and a failure. When the program failed, like the Challenger explosion, many of the factors contributing to the failure had been traced to not following the plan, and yet the stellar suc-

Implementing the Solution

cess is attributable to the planning and testing that NASA does, as in the Apollo 13 mission's safe return. Even in following a plan, you cannot guarantee success, but you minimize your chance of failure, and you greatly increase the understanding of what you did so you can improve upon it later on.

Implementing ISDN is not as complex as launching a spaceship, but don't tell that to the guy who is in the middle of a bad installation. When things go bad, tempers will flare, problems will be blown out of proportion, and your project will start down a spiral that can collapse the project if you're not careful. Planning to test the entire solution and ease it out to your organization will give you a chance to learn this new technology and to deal with all of the organizational issues and procedures necessary to keep this technology working at its best.

The testing plan you will use for your project is really no different from the steps to be followed in any other new system implementation project, but the concepts are well worth reviewing, and here you will find the material adapted to the specific issues and needs of ISDN activities. As a reminder, the objective of your testing process is not to test every possible error but to cover enough situations to make yourself reasonably prepared and to adjust your design and plans for the reasonable events you anticipate. If you make your plan so comprehensive as to test every possible failure, you will never finish your testing and the project will never be implemented. TABLE 6.4 lists a number of the elements that should be in your tests, identified by project phase.

Throughout your testing process take notes, preferably in a notebook, spiral bound, and keep this book for the life of the project. Just as you were taught in labs during science class, the key to determining what you did and how to repeat or avoid that result, comes from the quality of the notes you take during the process. Not only will your life become better from this process, but the people you are supporting and your support resources will

TABLE 6.4 Project tests.

Phase	Test	Description
Development	Line	Remain with the phone company to validate the final installation of your circuits. This test will be concurrent with the installation of the equipment by you or your vendor.
		Make every attempt to use the line in each mode you have ordered. This will include basic voice, data, and packet-data calls as required. Use your service providers or equipment manufacturers as resources to make this happen. Don't let the installation technician leave until you know the line is working with your equipment.
	Equipment	Concurrent with having a new line, or better yet, if you have a line installed somewhere already, do a fully functional test of the equipment before the new line is installed. Check all of the parameter settings, connect the equipment to the devices that will be serviced by your TA, ensure that you bring the new equipment to a known working state.
	Software	For any software you purchase or develop, create a test plan that will walk the software through all of its basic functions. Begin with just the basic functions, then build out until you have fully exercised the programs for what you intend to do.
		As an added test, deliberately do things that you would never dream the user would do, like turning a PC off in the middle of a firmware download to a TA, or interrupting a file transfer by rebooting their computer — these things never happen in real life!
	Functionality	Test for all of the services the user expects from the solution.
	Procedures	Take the procedures you develop and try them from the beginning, as if you haven't ever seen them before. Keep a log of what you did, how you handled exceptions and gray areas. Use these notes to improve and correct your material, then pass the revisions to someone else and do the same thing over again.

IMPLEMENTING THE SOLUTION

TABLE 6.4 Project tests. (Continued)

Phase	Test	Description
Beta	Installation	Follow the procedures and walk through the entire process with equipment and lines starting from a *freshly* installed condition. From a practical perspective, this won't happen, because you may beta test at your alpha location; if so, fake it and imagine you are doing this from scratch.
	Functionality	Have the users put the software through complete functional tests. Either build out from simple functions, or have the user dive straight in and see if everything operates as expected. Then troubleshoot from questions or issues that arise.
	Training	Determine the skill set and competency that a user should have after your training program. Then test for this at the class and several weeks later. Evaluate feedback from the participants as to what should be changed to improve the training based on their use of the solution.
	Support	Put your support process through the ringer during the beta phase. Make sure you have all the procedures and systems in place to make this implementation as easy as possible for the users and the implementation teams.

thank you, because it will give them better information and help you to resolve issues faster.

Always go into a test with a specific set of objectives and complete the test with a postmortem to determine if you met your objectives. From there you can decide what you will do next, move forward, or revise and redo your test based on changes to the solution. Have the key people in the project sign off for every completed test that met its objectives. This will keep the test team in good graces with everybody. They were all informed, and it will resolve questions later on in a project if someone suddenly complains that something wasn't properly tested and that the development group didn't do their jobs.

■ MAINTAINING THE HEALTH OF YOUR SOLUTION

Now that it is in and you're all finished, nothing else to do but enjoy the benefits and kick back to watch a game or good movie. Sure, but only if you take a little time to make sure the solution keeps operating the way it should. Little things that can go wrong, probably will, and unless you have a way to discover these trends, you may find yourself in the middle of a problem that is out of hand, and no easy way to fix it.

The most difficult aspect of projects like this is keeping straight what equipment, procedure, or person is doing what, and where a fault may actually lie. For example, if you have a TA intermittently give you a status that the ISDN line has gone bad, what are you going to do to solve the problem? After all, the problem is most likely the line, because that TA is brand new, or has been working fine since you installed it. It could be a problem with the TA, the software you are running, or the telephone CO switch. By tracking the time and circumstances around the problem, you can isolate and more effectively troubleshoot the problem.

A recent personal experience, with the Motorola BitSURFR Pro, brought home this idea of tracking your problems to isolate what's wrong. I began to notice that every once in a while my ISDN line went away, the unit would flash red on what should have been a green indicator. The line would come back right away, but this was odd. I found the problem happening every few days, didn't seem regular at the time.

I notified the phone company, who promptly tested the then working line, and of course, they found nothing wrong. But the technician did say he had read a note indicating that there was a firmware upgrade for my unit, to fix a reset problem where the unit would reset once every 24 hours — a symptom like mine, but not identical. So I watched the line twenty four hours later, it failed at the right time. So I downloaded the fix, installed it a few minutes later, and my problem went away.

If it is possible, and many times people will resist this, have your users fill out a *simple* form describing their failure. Use E-mail if you have it. Give them a template to fill in what happened; let a program or the template fill in all the other details. Then put all of these reports into a central database and track them through resolution.

IMPLEMENTING THE SOLUTION

When you can, at least once a month, look into the database and gauge where most of the support issues are coming from. Try to see if any trends emerge, like a piece of equipment, a line, an area of the country, a time of the year, a group of users, a particular application, etcetera. These trends are key to improving your solution, discovering where to place responsibility for the failures and how to correct the cause, as well as developing proactive responses to new issues.

When you have a problem, go after all the help you can to solve this problem. With so much of the industry wanting this technology to be a success, you have a lot of friends ready to step in to help you resolve your problems. The help desks of the hardware and software folks are great, but you're going to find that many of the carriers also have set up special support groups to be aware and capable of troubleshooting your basic connection infrastructure, and identifying if you have an ISDN or application problem.

■ WHAT'S NEXT

Applications, applications, applications. Time to look at all of this material put together, with some innovative ideas on how to use ISDN and some stories from the road. The next chapter covers the telephones applications, so read on, install a line, and give a friend a call on your new ISDN line.

CHAPTER 7

Presenting the Telephone Applications

You're here, finally, at the first of the applications chapters. It's time to begin looking at the various ways you can use ISDN to improve your life through high speed communications. In this chapter, I will show a number of telephone applications, just how the use of ISDN can provide you with sophisticated telephone services, the kind that only large PBX users with sophisticated systems have access to.

The chapter starts with a quick primer on telephone services, to give you the background to understand where ISDN contributes to these services. Then you will move on to the BRI applications, in both the office and the home. After that you'll see how the PRI services can contribute to large phone systems.

■ TELEPHONE SERVICES PRIMER

There are a number of services offered by the phone companies, and many are marketed under brand names to distinguish them from everybody else's services, but they are the same none the less. What I will do is provide a review of these services, under the most common names I am familiar with. If the services are offered in your area under another name, the phone company can clarify that for you. Also, in Appendix B I have tried to list the major brand names for the telephone services in the United States.

Inbound and Outbound

The first distinction the telephone industry makes on calling services is whether they are inbound, outbound or both. When you purchase services, they are committed to one function only and nothing else. Because of the way these services work, you often find yourself limited in what you are able to do.

The telephone companies developed a solution to extend the functionality of a circuit, mainly for inbound calling. By routing calls directed to a specific number (like an 800 number directed to one of your CO trunks), that call can be redirected to one of your inbound lines, giving it two separate functions, calls. If you have 5 lines that ring for the phone number 555-1212, you can add an 800 service to that line and the 800 line service provider would direct all of those calls to your 555-1212 line. But the real killer is that if a circuit has been designated as inbound only, then no outbound calls can be made on that line. Look at the next Sample Application story for a more specific example of this problem.

This example seems simple enough, and it really is. It demonstrates that ISDN's ability to control the use of each B channel gives you a flexibility in your calling patterns that in previous years was much harder to accomplish, or even do.

Dedicated Services

Many of the most powerful services offered by the phone company require dedicated lines to deliver the service. It is this use of dedicated lines that allows the phone company to route and handle the calls properly, and to offer you some control over the

PRESENTING THE TELEPHONE APPLICATIONS

> You want to have some DID trunks to allow a limited, but guaranteed, number of inbound calls to your staff. And you have done the same for your outward dialing needs. For an analog service you specify how many lines you want, and that number of dedicated lines are installed. Over time you find that your staff is starting to relay complaints that your customers and associates are getting more busy signals when they try to call. You look at how many lines you have installed, and you see 20 lines for 10 people, that must not be the problem, right?
>
> Here's where part of the problem resides, you installed 12 lines for outward dialing and 8 for DID. When those outward dial lines are idle, they are not available for any other purpose. The same holds true for the DID lines, if you had more than 12 calls to dial out, you wouldn't be able to use the DID lines.
>
> This problem has been a classic load-balancing issue that telecommunications companies and managers have tackled for decades. And this is where ISDN helps out. ISDN allows for the dynamic allocation of lines to a given call type. So if you have any lines available in an ISDN circuit, your call is automatically allocated space on a line without any effort or reconfiguration by you.

level of cost associated with these services. Here are some of the services that fit into the category of dedicated services.

- 800 — This service provides for toll-free calling into your facility. Originally used by major companies for large quantities of calls, the industry has brought this service down to a level of personal 800 numbers. Usually these calls are routed down the first available line in the dedicated line group, but with personal 800 numbers, the calls can be directed to any phone line supporting inbound calling services.

- 900 — A pay as you go service, these lines allow you to charge the caller for the call, but again on an inbound basis only. Through these lines, you can bill for information you provide, like support services, stock quotes, or an astrological forecast. These calls are routed down the first available line in the dedicated group and billing begins some time after the call is answered.

- WATS — An outbound dedicated calling service that provides for discounted calling rates based on the volume of calls on these dedicated lines. Calls go out on a line selected by the caller or by assignment of the phone system; if all lines are busy, the call is denied.

- DID — Direct Inward Dial is a dedicated line service that gives you a bank of phone numbers that are assigned to individual stations at your site. When a call is directed to one of these DID numbers, it is routed down the first open line, and then your telephone system completes the routing to the correct station.

- DOD — Direct Outward Dial is a dedicated calling service that commits all calling traffic to outgoing calls only. This type of service provides for a committed availability of lines for this purpose. If all lines are busy, then no other calls can be made unless an alternative line arrangement has been made.

- Centrex — This service provides the benefits and features of a phone system by using the equipment at the central office instead of your own. You purchase or lease the phones only, the phone company provides the rest of the equipment. Like Xerox, whose name is synonymous with photocopying, Centrex is the brand name of Pacific Bell for these services. However, if you refer to this name, most phone companies will recognize what service you are asking for. I will spend more time on this service in the next section, because of the features and benefits it can provide you.

As you see, there are a number of services available, and this list barely scratches the surface of what the phone companies have assembled for you under many different names and combinations. But it is the use of dedicated lines to deliver these services, and the fact that you must predict your calling patterns and install a fixed number of lines up-front that contributes to many situations where callers have become upset with telephone services. If your calling patterns go outside the original assumptions you used to determine how many lines were needed, you will have instances of missed calls and complaints, or lost business from those people trying to reach you.

Centrex

Centrex is a service that the phone companies have been offering for many years to relieve you of the responsibilities of

PRESENTING THE TELEPHONE APPLICATIONS

owning a phone system. Through these services you are able to perform many of the functions that a PBX provides, without the expense of purchasing a system, or the long term expense of keeping it up to date and operational.

This strategy has created a powerful set of features that has lead to many PBX owners to purchase Centrex services in addition to their own equipment. This happens because many of the features included in a Centrex tariff are charged for separately with other tariffs. Examples are features like hunt groups and call transfers off premise (check out the next bright idea). Hunt groups allow you to have a group of lines service a call type; calls on an 800 number can be directed to a main line, which if busy, will hunt to the next open line in the group.

> Under Centrex it is possible to transfer a phone call to a location out of your office. When the call is received by your receptionist, and it happens to be for someone who's working at home, the call can be transferred to him or her, freeing up the original line the call came in on, and letting the caller think that the employee was in, even if she or he is taking the call out poolside.
>
> And if your employee lives close enough to the office to make the rate affordable, you can actually extend your Centrex service to include her or his home. In doing so, two things happen. First, the call can be handled as if you were in the office; you can place the call on hold, consult with someone in the office, transfer the call back, all without your caller knowing that you were conducting business in your pajamas.
>
> The second is the fixed cost this line will run you. When a line is Centrex, all calls between the other lines in the Centrex arrangement occur without per minute charges.

Another benefit of the Centrex features are the remote office implications. Usually, you can configure multiple locations, even outside your CO, to be a part of your Centrex group. This provides a consistent phone functionality across all locations. The call transfer features give your caller the feeling that everybody is located in the same office. This type of functionality has only been available to PBX users with private phone networks. And the cost for these services is based on the local calling cost, plus any mileage charges if the remote office is outside your CO. (Note: some tariffs allow this functionality for analog Centrex lines, but not ISDN, so check for the availability of this service with your phone company.)

■ ISDN BENEFITS

It is the ability of ISDN to provide different services on every call that gives you a way to break the service barriers that traditional services impose. Like the first example, with a PRI circuit, you can deliver DID, WATS, 800, and 900 services all down the one circuit, plus have Centrex services on top of that.

Understand that these services have to be offered by the same carrier for you to realize full cost benefits. If your 800 service is delivered to you by a long distance carrier, your WATS is from your local exchange company, and your 900 service comes from one more carrier, then there would need to be three sets of dedicated circuits, one from each carrier — there go the benefits of ISDN — almost.

When ISDN is used for one of these services, new capabilities are made available to you providing for better levels of services. So even if you get caught in the dilemma of multiple circuits to different carriers, you may still decide to go with ISDN for these other features. In TABLE 7.1 is a sample of features provided by some of the major services as a result of ISDN carrying that service, and some of the overall benefits that an ISDN circuit provides.

With a BRI the line provides you PBX and Key Systems benefits. From individual phones it is possible to have a complete operator's console, with all lines in an office present on the phone with indicators on the line being used, etcetera. As calls come into this console, the operator can answer them, place them on hold to handle other traffic, page the person being called, route the call to voice mail, or any number of other phone system functions.

PRESENTING THE TELEPHONE APPLICATIONS

This capability brings to offices, from two to hundreds of people, new choices for phone services. And when you consider that these same features can exist on almost any other phone in the office, your imagination becomes just about the only limit to what you can do.

TABLE 7.1   Additional ISDN benefits for selected services.

Service	Benefit
All	The call setup time is typically 300 milliseconds in your local area using ISDN, compared with 3–7 seconds for analog calls. This benefit alone can save you tens of thousands of dollars each year.
	Call-by-call services. Going beyond dedicated lines for each service, ISDN gives the ability to use each line for any of the services you have subscribed to from the carrier supplying your circuit.
	The ability to carry data across your lines, as either ISDN packets between ISDN TAs or as X.25 packets, going out through the public-packet network.
	ISDN provides a full 64 Kbps data rate because of the D channel messaging services. This allows for high quality audio and data services across an ISDN circuit.
800	Automatic Number Identification (ANI) provides the caller's 10-digit phone number, which can be routed to a computer system and used to trigger both the proper routing of the call and to bring information up on a computer screen when the call is answered.
	It is possible to reroute calls to alternative locations by use of the messaging capabilities of ISDN, extending the normal line failure feature to a call overload situation.
Private Networks	ISDN allows the transmission of user-defined messages and other call information through the D channel, providing the appearance that remote PBX systems operate as a single entity.
900	It is possible to change the billing rate of a 900 call during the call. This is useful for calls that start as billable and are determined to be no-charge, like support calls.

BRI was built to deliver many voice services, including: conference calling, call forwarding services, call hold, multiple call appearances, and others. The full range of services that will be available to you are dependent on your telephone company and the current level of ISDN they have on their CO switches. But enough said about the promise, let's look at some specific implementations of ISDN voice services. We'll start in the home, a telecommuter's dream, and move our way out to the office and a full-blown PRI implementation.

■ THE APPLICATIONS

ISDN to the Home.

There are a number of benefits that arise from an ISDN phone service in the home. Many of these are derived from the mixed data and voice capabilities. For now let's look at just the voice services and see what they can do for you.

In many homes today you find two or three pairs of wires, allowing no more phone lines than you have wire pairs. But with the demands of today — a line for the family, a line for one spouse's work, perhaps a second line for the other spouse, and then the kids grow up to be teenagers and want their own lines — you flat run out of lines, what can you do?

Enter ISDN, where a line is no longer a line. With ISDN you can use the wire in your house more creatively, changing the whole character of your phone services. Let's look at the transformation of a home with three lines into one with four lines and a very creative set of voice services. The original wiring diagram is shown in Figure 7.1, where there are three wires routed throughout the house, and I am going to show you how these three wires can become up to six distinct lines, with a little ingenuity.

One of the primary problems with in-home wiring is the way the original analog lines worked. All the phones could be connected in a serial manner, meaning that one wire pair could be run from outlet to outlet, in series, and all the equipment plugged into the one wire. Whenever two people pick up the line, they both hear the conversation and can participate on the call. ISDN phones don't allow this type of wiring, your first stumbling block to reus-

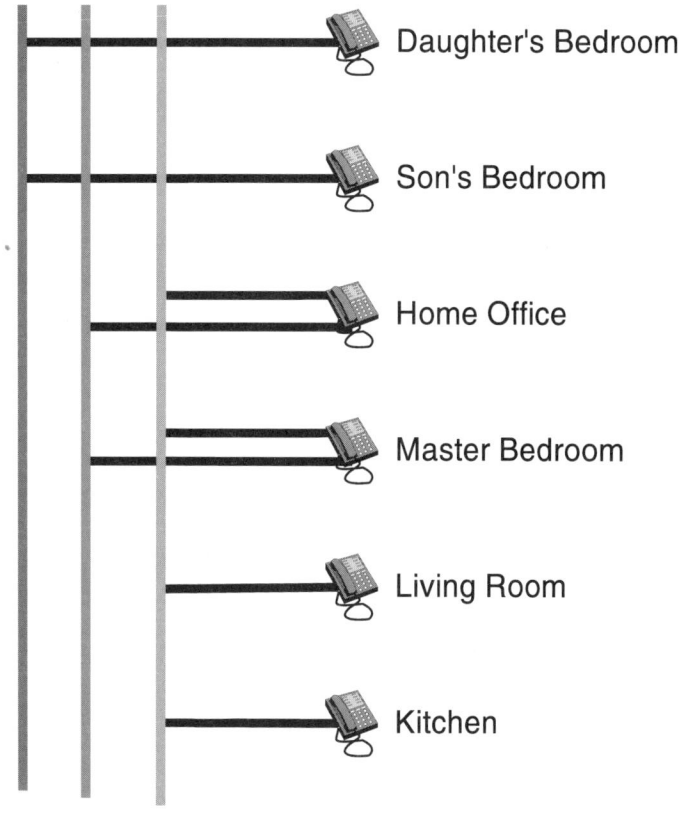

FIGURE 7.1 HOME WIRING DIAGRAM — LOGICAL FORMAT.

ing the existing wire. Also, the wiring usually doesn't run from room to room in a logical order, it follows the path of least resistance at the time the house was wired. And that may make no sense once the house is finished, especially if the house is multiple stories. Figure 7.2 is a layout of how the wiring might actually run.

But stumbling blocks are not barriers, and this one is certainly solvable. If you are fortunate enough to have the time, initiative, and persistence to see the problem through, you can determine the exact sequence of your outlets. And by knowing this sequence, you can solve your wiring plan and do what I did in Figure 7.3. I staged the TA devices so they reused the wire in each

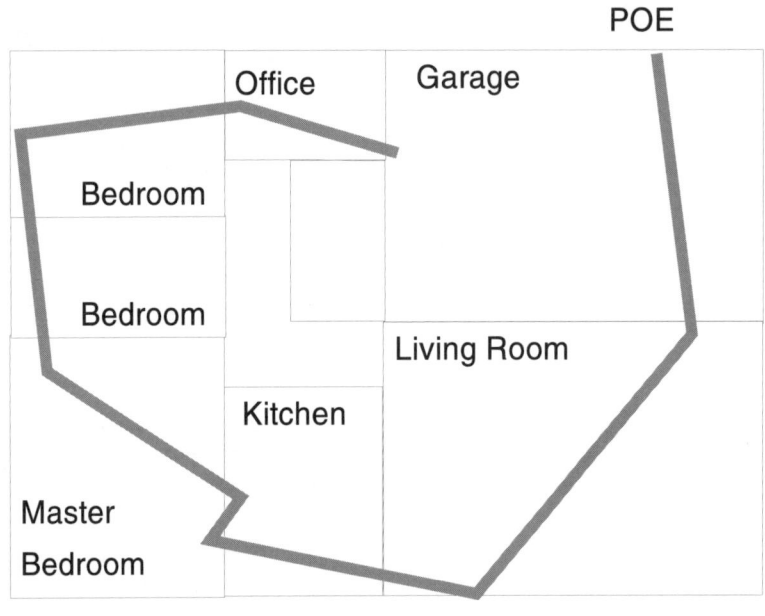

FIGURE 7.2 TYPICAL HOME WIRING ROUTE.

area of the house in such a way that I could get six phones with four unique numbers, out of the three inside wires.

Remember that you must have an NT1 terminate the ISDN line from the phone company, and that will be the U-interface as required. After the U-interface you can only run analog or S/T lines, and an S/T line requires four wires. So my choice in this situation is to reuse the analog phones and just use a TA with an NT1, two analog ports, and an S/T port, to split each line into two analog circuits, still providing me at least one S/T interface for other uses (like a data device).

I start by interrupting the wiring of the house, enabling me to break a single wire into two, as shown in Figure 7.4. This technique allows placement of a phone in each room with the TA, and then sending on the second analog phone line to other rooms down the wire that carried the ISDN line to the TA. Since each analog line carries a unique directory number, only the phones connected

PRESENTING THE TELEPHONE APPLICATIONS

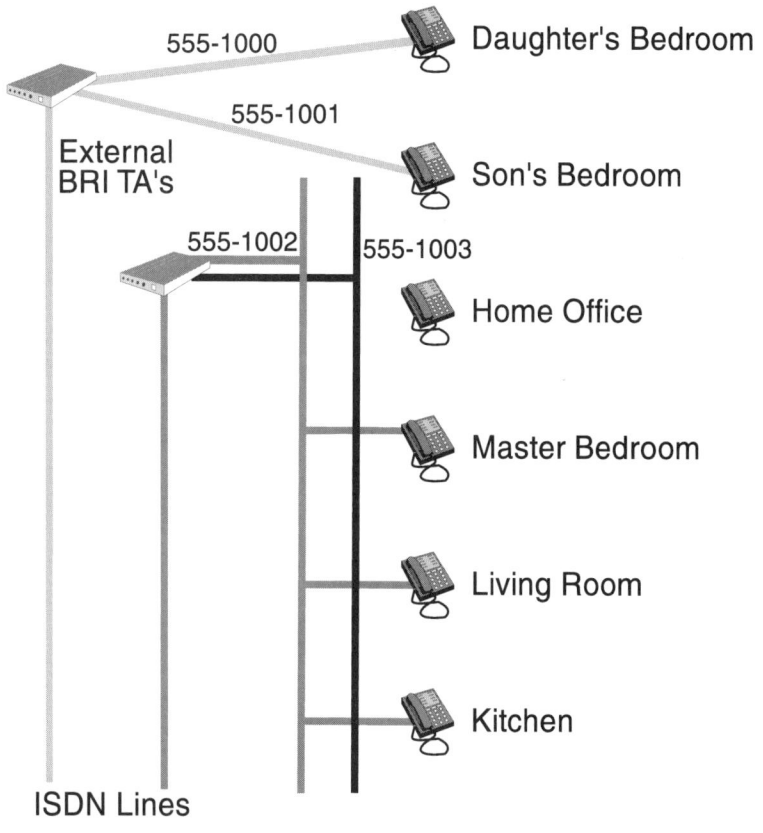

FIGURE 7.3 ISDN WIRING PLAN FOR ALL ANALOG PHONES.

to that directory number will ring, allowing you to have four different numbers from the two BRI lines.

While the equipment cost in this situation may seem expensive at first, this expense is actually a bargain, when you consider just the headaches of wiring more phones; waiting for a line to free when your teenager has a hot call; the family grief when you need the phone and so do they; plus the other things you can do with your time and with ISDN in the data applications I'll cover in the next chapter.

The real cost here will be the two TAs you have to buy, at a cost of about $350 each. This solution, without your time to

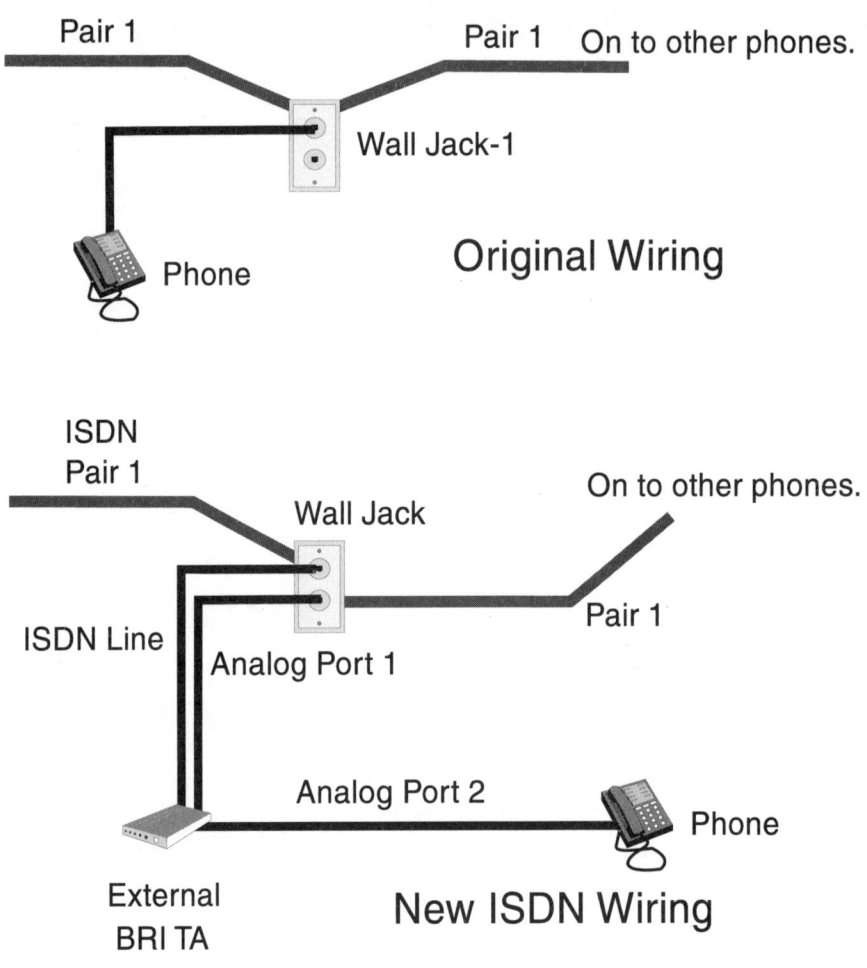

FIGURE 7.4 SPLITTING A SINGLE LINE INTO AN ISDN TRUNK LINE AND TWO ANALOG LINES.

rewire or the installation cost of ISDN, will cost about $700. Since installation costs are variable and often times are waived, it's hard to predict these costs. In California, right now, new residential lines are $70, while conversions of existing home lines are free. The ISDN line installation charge, separate from the normal installation charges, is waived, saving $125 for each line. And the monthly recurring costs will be about $25 per ISDN line, or $50 for these two circuits, with four analog lines. Normal line costs for four circuits is $44, so you have a slight increase in the monthly cost.

Another twist to this application story is changing the lines over to Centrex. Your cost jumps to $64 per month, but you get call transfer and rollover capabilities. Even with analog phones it is possible to use these functions, giving you a fully functional phone system in the house. You could be on the phone, and have another call roll over to a second line, perhaps to an answering machine or a voice mail service. You could also conference call, having people on all four lines in the house talking at once. And with the ISDN phones, it would be possible to give each person multiple lines for his or her own use.

This last suggestion is interesting, because it gives you the ability to have two, three, or more lines per person with an ISDN phone. And that is a powerful tool when dealing with anyone's need to expand her or his phone capabilities.

Small Office

There are many advantages in an office building that aren't found in the home. The primary one is the way the telephone wire has been installed. In office buildings, the wire always starts in a central wiring closet, where the phone company has installed punch down blocks to help them cross connect to a wire that runs out to one location in the building. It's this single wire facility that makes wiring an office for ISDN an easier task and gives you more options.

Since we have seen how the BRI lines can be split by the NT1 devices, let's jump into a small office design and dissect how it is implemented. In Figure 7.5 you want to focus your attention on the way the wire closet has been setup. Figure 7.6 shows the distribution of the ISDN and analog circuits from the wiring closet out to the different phones, faxes, and modems.

■ Wire Closet

> **110 Block — the AT&T product for Category 5 cable punch downs.**

In the wire closet you can see that the initial delivery of ISDN is on two wires, terminating on a 110 block, allowing you to route the line to any location in the facility. You can convert the ISDN line to four wires at this point, as I have in this example, or you can carry it forward on two wires to the final destination. This arrangement, Figure 7.5, works well for distributing both ISDN and analog circuits from a central point.

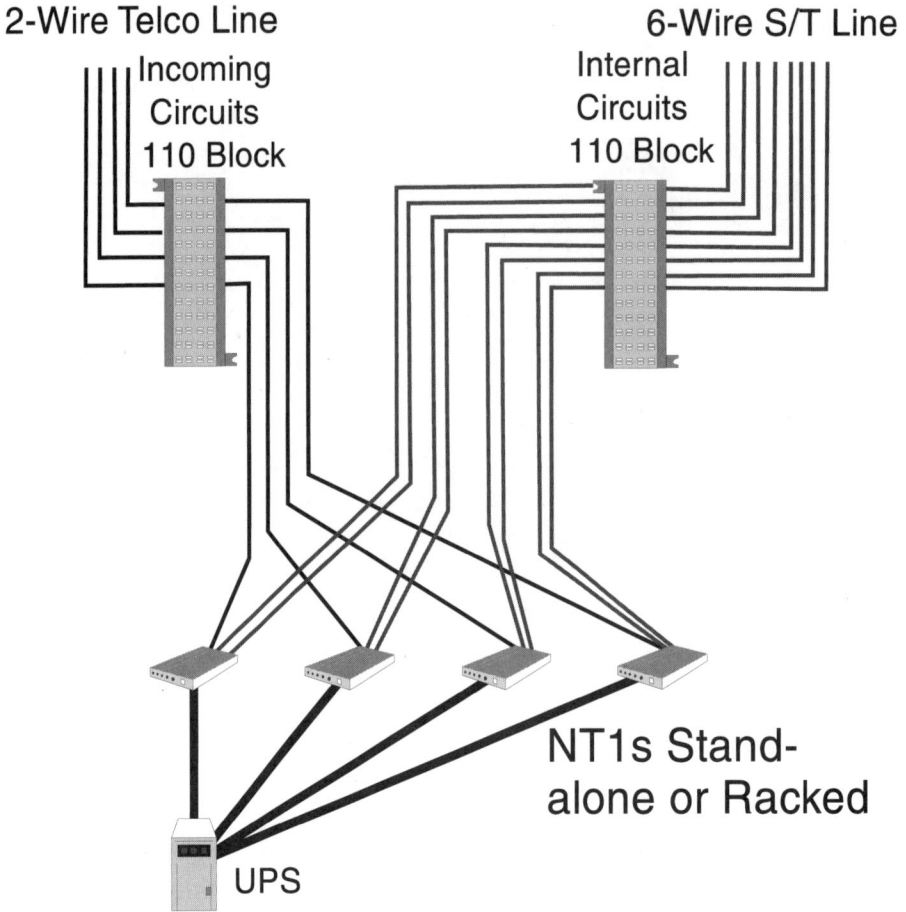

FIGURE 7.5 WIRE CLOSET CONFIGURATION FOR SMALL OFFICE USING BRI WITH CENTRALLY SUPPLIED POWER FOR ALL CIRCUITS.

When you are distributing from a central point like this, it is easier to control your troubleshooting efforts, and you can provide for two of your power issues. The first is the power required to operate the NT1 and ISDN phones. The NT1 can provide this power for each phone down one spare pair of wires going to the phone set, reducing the amount of equipment at the user's desk. By sending the power from a central source, you can put a single UPS in to provide battery backup for continued phone service during a blackout.

Did you notice that each NT1 split the ISDN circuit into two lines? This is a bit of circuit sharing you can do, so each person can be assigned a single B channel with a unique phone number. If you have ten people in the company, you can purchase five BRI lines to give each person his or her own phone. If there is no need for everybody to be on the phone at the same time, you can increase the number of phones on each terminal adapter to as many as eight, but you will be limited to only two calls at a time in this arrangement. Two other things have to happen to allow you to go beyond two phones per line today. The first is an NT1 with a passive bus supporting that many phones, and second, support from your CO switch for that number of devices. National ISDN-2 will provide that support, but that software and hardware is not yet available everywhere.

■ Office Setup

Let's look at how some of the users are set up for using ISDN at their desks. Figure 7.6 shows a few different office arrangements: the first is just an ISDN telephone; the second has an ISDN telephone and a fax machine; while the third uses a fax modem, fax machine, and an ISDN phone. In each of these situations, Centrex services route the calls to the appropriate desk or device, in addition to the ISDN features that handle some of these duties.

In the single phone office, you can install more than just one line or simple calling features. Make this a speaker phone to start, add the conference calling features of ISDN, and now you can

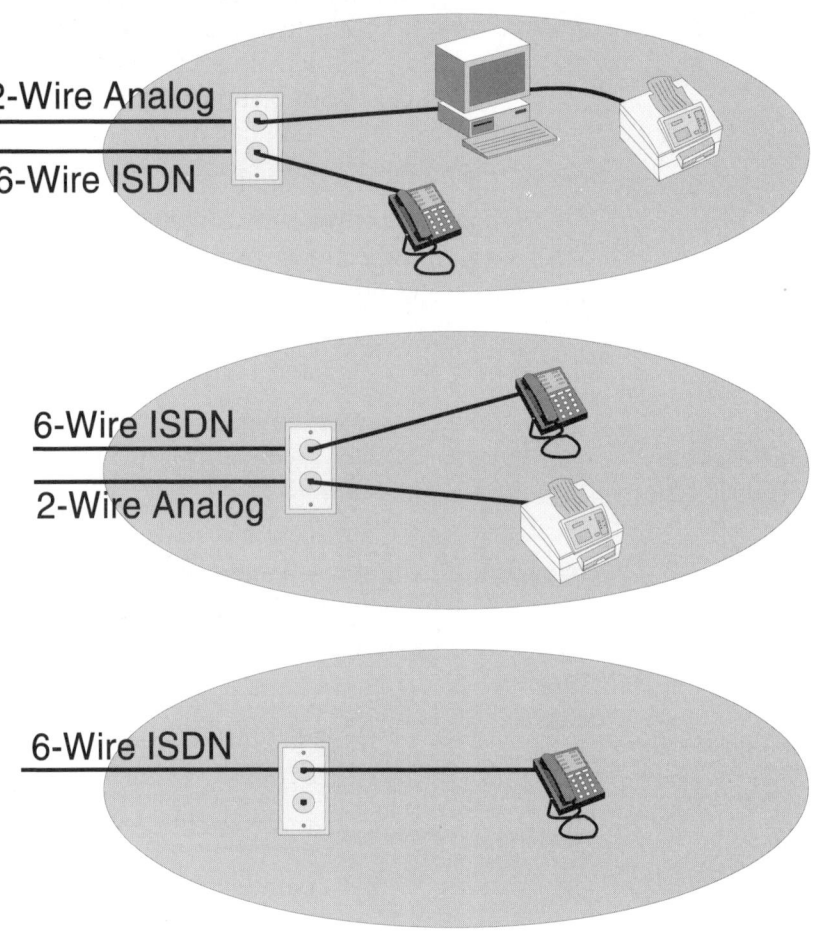

FIGURE 7.6 THREE OFFICE CONNECTION EXAMPLES: SINGLE PHONE; PHONE AND FAX; PHONE, FAX, AND COMPUTER.

carry on a call with two other people down this line. Move to the advanced ISDN EKTS conference calling features, and you can extend this conference capability to six people.

With a few feature keys on the phone, you can configure additional call appearances. Several options are available:

- Other phone numbers in the office can appear on these keys, allowing you to answer their lines, join in on their calls, and basically provide phone coverage for these folks.

> **Feature Keys —** all those nice buttons on your phone, which are programmable on ISDN sets and phone systems.

- You can configure additional line appearances for your own line. If a second or third call comes into your phone number, you will see the call flash and/or ring on these additional keys, giving you the option to answer the call. You would answer the call by placing the first call on hold, which frees up the B channel to transmit the second call.

This line can also be configured to roll over to another station after so many rings or if your line is busy (when you're juggling three calls already on the multiple appearances). This rollover could also go to voice mail, or even to an analog line with an answering machine.

In the second office example, you are bringing in an analog line to service the fax (or an ISDN line for a group 4 fax machine), and you have one ISDN line for the phone. All of the phone features above are going to apply again, so what's different? The analog service is being delivered out of an analog port on the NT1, taking a single B channel and making it a voice grade line to support the fax services.

If you get creative and have a need for multiple outbound fax ports, you can route them through a single B channel if you don't mind a little traffic congestion. This works well if you have two or three people who occasionally need to fax things and don't mind waiting a few minutes if the line is tied up. Just wire the one fax B channel to each machine, splitting the line at the 110 block. In this arrangement, you need to select one machine for the inbound faxes to avoid the machines fighting each other for control of an incoming fax. This arrangement works best when you have a centralized person responsible for all incoming faxes, and you want to let everybody transmit their own work.

In the third office, all that you did was to add the computer in the middle between the wall jack and the fax machine. Both machines know how to work nicely with each other. Just make sure that only one of them has the answering responsibilities; otherwise, they will spend a lot of time fighting for control and your inbound faxes will drop into never-never land.

■ Home Office

There's not a lot of things better in life than being able to work from home and have all of the conveniences of a larger office. After all, being able to get up minutes before a conference call, or sleep in that extra hour, then take your one minute commute and be ready to go makes it all the better. Let's look at the configuration of a slightly unusual office that integrates ISDN into a home and key system environment, with mixed application uses.

Figure 7.7 shows the way this solution has been configured. ISDN came into this situation after everything else, so the challenge was working in features that make the most of adding another line. The need to add a fourth line forced the move to ISDN because the house had only three lines in the walls, and the office functions were tripping all over each other. In this situation, the two businesses, a bookkeeping practice, and a consulting practice, had one business line each, and the home line tripled as a personal, fax, and a data line. As business was good, and many personal commitments were being made, the third line was consistently busy when it was needed for something else.

Voice mail at the time was going off-site to a phone company service, so each business line would roll over when the phone was busy or not answered. This was working fine, but fate provided for a phone system with voice mail to be acquired. This forced a change to the services, and now the home line had another duty added to it — rollover responsibility to the auto-attendant and voice mail system.

That was the straw that broke the camel's back. Something had to be done, and ISDN was the least expensive path to a solution. At the time, the tariffs allowed for the conversion of home lines to ISDN at no charge, which took care of one major expense. Next came the cost of the TA, which was a Motorola BitSURFR Pro because of the support for two analog ports. Prices on that product were in the high $300s and have since dropped to the mid $300s. Beyond that was the monthly cost of the line being added,

PRESENTING THE TELEPHONE APPLICATIONS

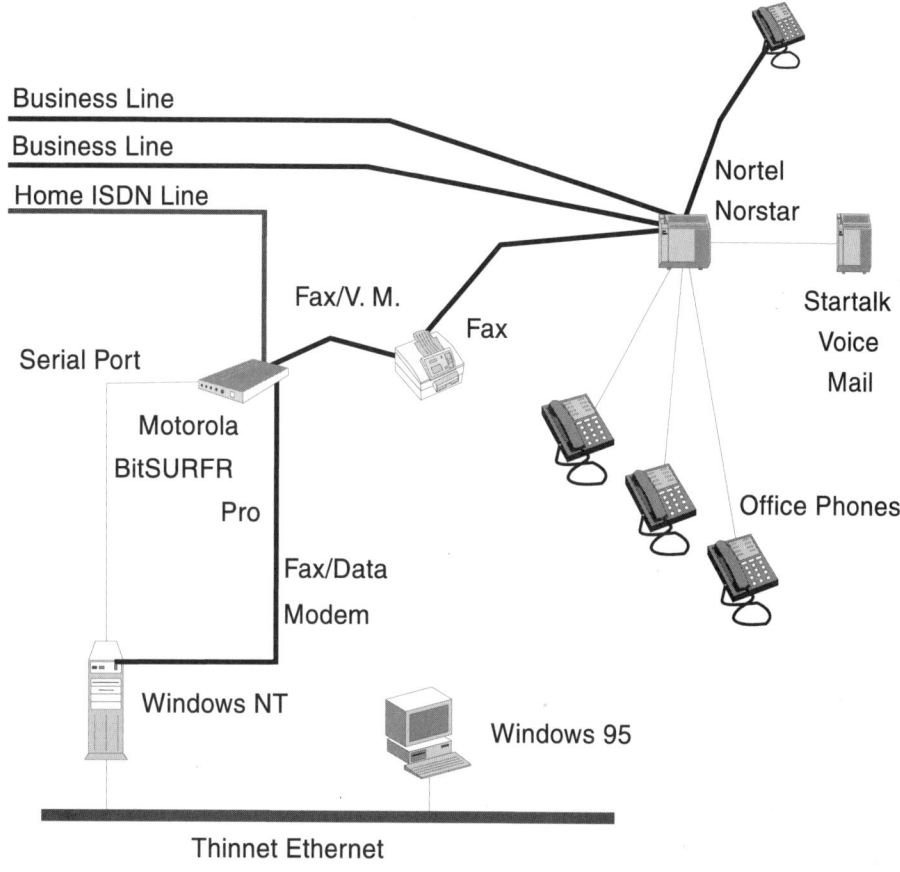

FIGURE 7.7 HOME OFFICE WITH DATA AND KEY SYSTEM APPLICATIONS SERVED BY ANALOG AND ISDN LINES.

instead of roughly $12 per month for the home line, the cost was $25, plus usage charges during the weekdays — an added expense that averages about $20 per month. So the net increase each month for just a fourth analog line was roughly $21.

So what was worth the extra money? First, no extra wiring inside the house. Second, up till the time the ISDN line was installed, the home line had been experiencing noise problems (this was static in the lines, affecting the fax and computer modem), which all but disappeared with the ISDN line. Third, the two B channels were configured for circuit-switched data, which allows for calls out to: clients, on-line services, and Internet providers at speeds up to 115.2 Kbps speeds, a far sight better than the 14.4 Kbps modem currently installed.

So what's ahead for the site? Figure 7.8 shows how the site will be configured with a second ISDN line. The primary purpose of the line is to fully isolate the fax and data services from the voice activities. As things currently work, the fax/home line still provides the rollover for each business to the voice mail system. The second ISDN line will isolate the voice services to the key system and the house phones. With the addition of a 900 MHz portable to the key system, access to all three voice lines will be possible throughout the house with no additional wiring.

The last challenge to conquer is the placement of phones throughout the house, connected to the phone system and the ISDN lines. To do this the existing wiring has to be traced. The house is two stories with an unusual floor plan, and almost twenty years old, so a long weekend, a few friends, some pizzas and beer, and maybe this task will get down. Once this is done, the creative wiring in Figure 7.3 can be used to put two lines up around the house to give six extensions off the phone system, enough for this family of four with two businesses in the home.

PBX-Based Offices

As you create a sharing arrangement for ISDN services at a larger office with a PBX, your options expand. You have to decide if it is worth the expense to bring in PRI lines into your PBX and split out ISDN lines through the phone system; or bring in BRI for specific applications and share the voice capabilities with your PBX. In this section I'll show you both variations and discuss some of the issues to resolve before you can make the decision.

One of the most exciting aspects of bringing ISDN into a PBX is the immediate benefits of PRI managing the circuits. Remember that you get call-by-call service selection with PRI, and that allows you to combine a number of services into one line;

Presenting the Telephone Applications

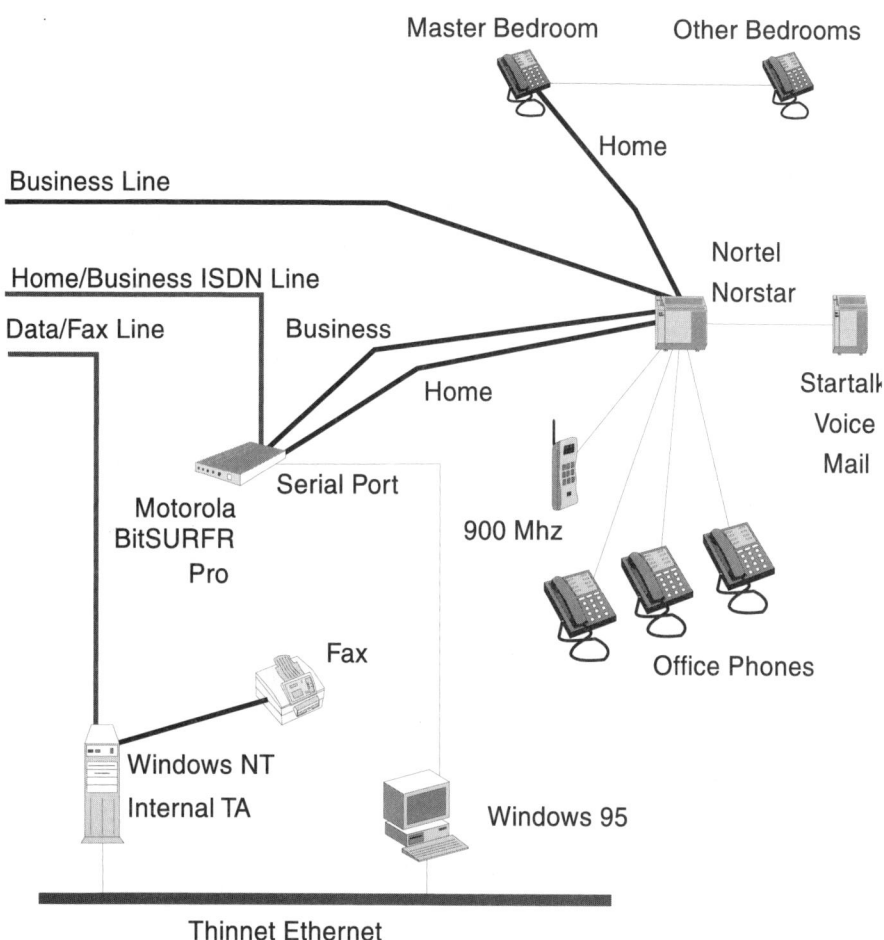

FIGURE 7.8 NEXT IMPLEMENTATION OF A SECOND ISDN LINE IN THIS HOME OFFICE EXAMPLE.

the objective being several things: reduced line cost through load balancing, better service through the load balancing again, and the delivery of new services that use the digital data features of ISDN.

There are three configurations that I will explore with you, but you can come up with many more than this, because the number of different ways you can interconnect this equipment seems to have no end. The first will be the PRI line coming into a PBX with a PRI interface, plus the ability of the PBX to distribute BRI lines for the various data and videoconferencing requirements in the facility. The second will look at the use of a Drop and Insert device to provide channel allocation to ISDN devices the PBX cannot serve. And the third will be the integration of BRI lines into the PBX to utilize circuits that remain idle during peak voice periods.

■ *PRI and the PBX*

Many of the newest PBX systems are capable of managing ISDN services. The most popular application of ISDN on this platform is the use of the call-by-call services. When your line capacity begins to reach the upper teens, you need to begin thinking about consolidating lines to carry all of that traffic. The traditional alternative has been the T-1 circuit, where you could get 24 channels allocated to specific services. The phone companies frequently offered a financial incentive to move to these services, because it reduced the number of copper pairs used in the distribution of telephone services, allowing them to add new customers and capacity over the existing cable without expensive cable upgrades or additions.

PRI goes a few steps further today in providing services to you, but the incentives to move to this technology have changed. The telephone companies have upgraded their wire facilities, so many of the financial incentives have disappeared. But the ability to transport both voice and data provides the greatest drawing power for PRI over any of the T-1 services your phone company has to offer.

In Figure 7.9 you see that interfacing to the PBX is relatively straight forward. In this application the PRI arrives on two wire pairs into an NT1 and then continues on four wires at the S-interface point and passed on to the NT2 and the PBX as 23 channels of voice/data and a single D channel. The PBX then proceeds to manage the traffic on the 23 channels according to the configuration of services and your PBX.

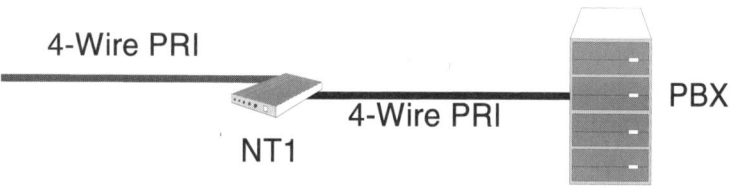

FIGURE 7.9 PRI INTERFACE TO A PBX.

What makes the implementation of PRI a little more difficult than BRI is the volume of traffic it can handle. In many medium-sized telephone system environments, you may have 10 to 20 lines, but these are spread out as local phone traffic, long distance traffic, and other specific applications. With the PRI you need to plan for how you are going to distribute those services across this one circuit. In fact, look at Figure 7.10, where the services are delivered on normal analog lines and contrast that with Figure 7.11 where all that traffic has been consolidated into two PRI circuits.

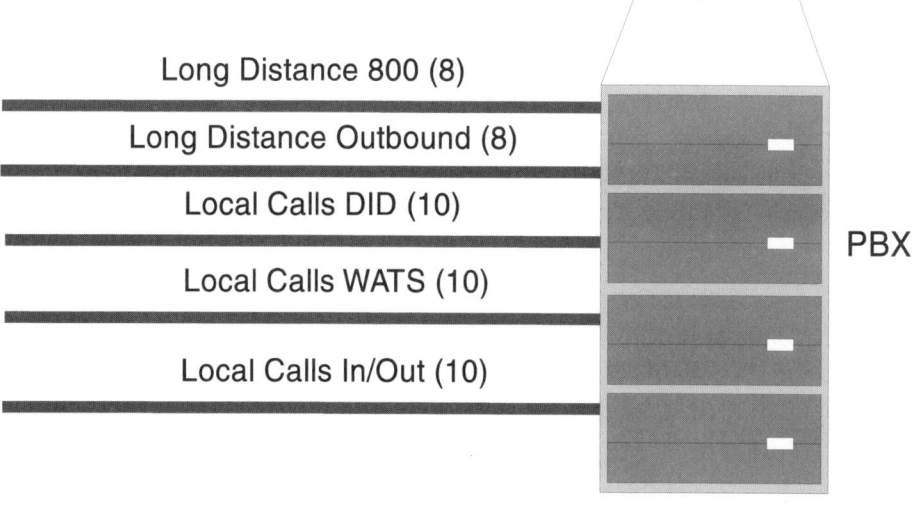

FIGURE 7.10 SAMPLE ANALOG LINE DELIVERY TO PBX.

The original line count between both sets of services was 46, and the replacement line count is also 46. The balance is different, since the local lines have dropped by seven, and the long distance

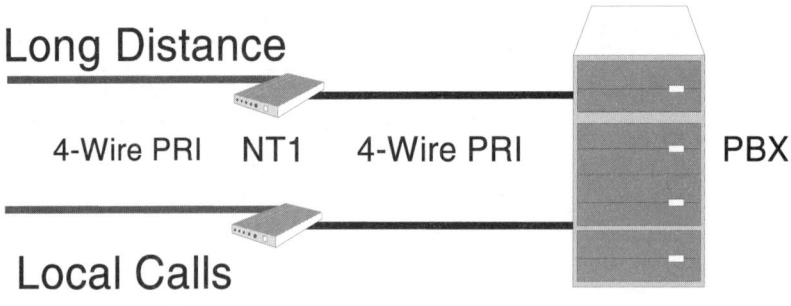

FIGURE 7.11   REPLACEMENT PBX SERVICES DELIVERED ON PRI.

lines picked up the difference. This is a good example where you must decide if your new calling patterns justify the new balance. In this situation, it probably will, because there is a good chance that the local lines have never been fully used, all 30 at once. If business is good, the long distance lines have been tied up more often than is comfortable. Since PRI will let you mix and match these services, any one of the original services could have more lines than they did before, by the benefit of another service being idle or less than full capacity.

This whole idea of mixing services on the PRI can lead to substantial cost savings. For an example, just the reduction in the time to handle a call can result in tangible savings and increased capacity. Let's consider these two instances:

> An 800 service that provides aluminum recycling center locator services implemented ISDN and reduced their call times by 12 seconds, yielding an average savings of 4 cents per call. This happened by connecting the ISDN data about the caller to a computer system that could retrieve the caller's area code and prefix, then matching that to a database to find the recycling center that covered that area.
>
> If you're doing 300,000 calls per year, this would return a tidy $12,000. Given that a national service like this only has to do less than 1,000 calls per day to reach this savings level, and recycling is still a popular activity, this is a significant opportunity.

PRESENTING THE TELEPHONE APPLICATIONS

> For another company, the speed of the calls being set up and handled affected their total call handling capacity. By implementing ISDN, the call center was able to increase their call volume by about 1,000 calls per week. Being able to respond to even a fraction of this total number goes a long distance towards customer satisfaction and market perception, and that is, at times, priceless.

Another powerful feature of the PRI ISDN technology is the ability to communicate through the D channel to link remote systems and services into what appears to be a single PBX and voice mail system. Remember that your ability to use the packet-switched data services is limited by the carrier being able to carry that traffic through their public-packet network.

Back in Figure 7.11, there is a PRI coming from the long distance carrier. If each of your sites has the same arrangement, then you could implement a sharing of voice mail information and functions, making your network look like one system. Another story from the trenches talks to the potential savings in this situation.

> The Audix voice mail system is a powerful and popular platform in the marketplace. To link these systems many companies have used private line services to provide for the fast response time required to make remote locations appear to be present at the same site where the original call was received. But private lines can be quite expensive, and because they are up all the time, you pay for bandwidth that goes wasted most of the day.
>
> In this example, two locations were located over 2,000 miles apart, and the use of ISDN reduced the annual costs by over $6,000. This is possible in part because the call setup times, measured in milliseconds, allows remote systems to interconnect quickly enough to make this remote service arrangement acceptable to the caller. That fast setup time leads to letting the line drop in between calls, releasing phone company facilities, and letting them charge you less for your services.
>
> Now don't think the phone companies are all that magnanimous. They know that if you are providing better levels of service, your business will grow, your need for services will grow, and hence, their business will grow with it. Your real benefit is more business for the same amount of money, and that should mean better overall profits.

In the next chapter about data applications I'll show you how the PBX can help you share PRI lines around the clock.

■ Drop and Insert

What if your PBX can't handle a PRI circuit, but you need to add some lines with ISDN capabilities anyway. As in the days of T-1, the drop and insert capable devices come to the rescue. What this device does is intercept the ISDN line and provide the channel management that your PBX and other equipment can't. It distributes the channels as either ISDN or analog lines as you need them.

You will receive the benefits of the ISDN technology where you need it in native form, and you will get the call-by-call services for the PBX, with the added expense of two sets of line interfaces — one in the drop and insert device, and the second in the PBX or other equipment. If you want to make moderately good use of an ISDN line and the reduction in monthly expenses offsets these costs, then this approach is much better than doing nothing.

Figure 7.12 provides a simple design for installing a drop and insert device. As you can see, several lines have been routed out to videoconferencing and data equipment, your real justification for putting in this device. These devices need ISDN for their operation. On the plus side, now you can consolidate all of the traffic onto a single line from the carrier, in this case long distance, but it could have been a local carrier, saving you money in access cost and aggregating the traffic. Later on, when you upgrade your PBX to handle the PRI technology, you can eliminate the drop and insert device and get back to the business of using all of the circuits all of the time.

■ BRI to the PBX

In the worst case, you can bring ISDN technology to your PBX through BRI lines. Perhaps this sounds harsh; it isn't meant to, but this level of service is a strong fit in small key systems and for specialized application at the PBX level. So worst case just has

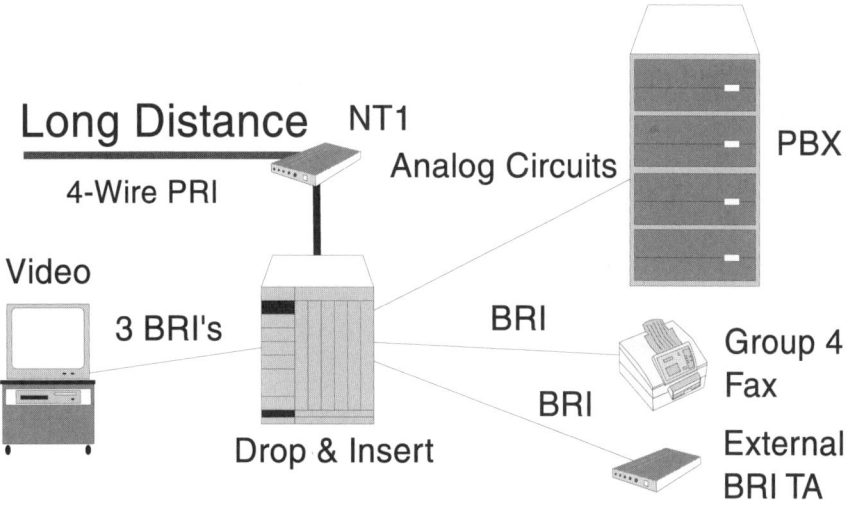

FIGURE 7.12 DROP AND INSERT EXAMPLE.

to do with the typical mismatch between the size of the switch and the circuit.

> The phone companies have contributed to the situation of installing a number of BRI circuits instead of a single PRI line. As an example, the SBC tariff for monthly rates in Texas is $48 per BRI circuit, $1,049 for PRI. Twelve BRI circuits would be $576, a savings of $473 per month. While initial equipment costs are higher, this $5,676 of savings per year can buy a lot of equipment, plus the safety factor that if a line goes out, only one part of the capacity is gone.

I won't go through the connection of a BRI directly into the PBX or key system, because it is the same as the PRI example. Instead, let's look at what I call the wanna be drop and insert device, the BRI TA with some analog ports. If you have an application that requires some ISDN services for devices that are outside the domain of your telephone system; but you hate the thought of that circuit lying idle for hours on end, then take the poor man's route to solving the problem. Take the analog ports of a TA and route those to your PBX or key system, much as the home office did in Figure 7.7. This approach will allow the phone system to use the analog services in contention with the data and

■ THE ISDN CONSULTANT ■

ISDN needs of the ISDN devices. You can look on this as a small bonus of capacity, giving you a little more breathing room before that next major phone system upgrade.

Final Example

An interesting thing is happening in the broadcast industry today. As you may know, many radio shows and audio events are being done from remote locations. In the past these folks have used wireless technologies or audio quality phone lines to get their broadcast back to the studio before it goes out over the airwaves. And in a fast response situation, the wireless approach still remains the best technology to use, but when the events are planned for, ISDN has stepped in as the effective and affordable way to get the job done.

It is possible, using both channels of the BRI circuit to reproduce sound with an almost CD level of quality. What has attracted the broadcast industry is the lowered cost of this service and faster turnaround time for setting it up. Today, there are circuits known as broadcast audio services, where the phone company will condition the lines and guarantee the transmission quality, all for a reasonable, but expensive sum. ISDN on the other hand, can transmit the audio signal in digital form, hence error free, over great distances, and at a much lower cost. So, who knows where that radio announcer really is, right there in your hometown, or a quarter of the way around the globe. What does it matter, the music's good, you like the programming, and it sounds great!

How are many of your favorite radio broadcasts really getting from the broadcast site? Two major sports activities, the NBA games, and the 1996 Olympic games, are using ISDN to distribute the radio broadcasts from the event to the central distribution points.

The Juilliard School of Music uses ISDN to link some of the world's foremost musicians to classrooms for lessons. And Zubin Mehta's recent concert was broadcast simultaneously in Los Angeles and Tel Aviv using ISDN as the audio link.

PRESENTING THE TELEPHONE APPLICATIONS

■ WHAT'S NEXT

Now that we have beat the bushes with many of the ways you could configure up ISDN in a voice environment, it's time to see what the data is up to, but don't be disappointed. The data side of the house is actually much simpler, because there are still just a limited number of ways you can configure this type of service — the creative stuff is yet to come. So move on ahead and have some fun.

CHAPTER

Presenting the Data Applications

When you're hot, you're hot. And the Internet and data applications for ISDN are HOT! All of the major media hype has gone into this area of ISDN. And all of the major competition from other technology focuses on the ability of ISDN to deliver data at high speeds.

The fact is that ISDN can deliver data at speeds 10 to 12 times faster than the old 14.4 Kbps modems that still dominate the market today. (Did you know, that as late as the latter part of 1995, over 50 percent of the people accessing the Internet were still using 14.4 Kbps modems, even though 28.8 Kbps modems have been available for several years.) Another fact, ISDN does more than just data, just in case the previous chapters haven't convinced you of that. If all you are looking for is a hot download technology, hold on for another year or so, 1997 should be a wild ride in pockets of the country and world, and 1998 should be hot — but for the rest of us who are dying to get a reasonable speed out of the Internet and our remote access applications, today we need to go to the ISDN table and suffer with 128 Kbps to 1.92 Mbps bandwidth.

■ DATACOMMUNICATIONS SERVICES PRIMER

I hesitated to simply describe this section as a data communications primer, because so much of the world seems to revolve around Internet access, and that is a networking service that uses a datacommunications access technology. But I know, as you do, that the world doesn't revolve around the Internet, the rest of our tasks have a life and this section will deal with those issues too.

Some History

The entire idea of datacommunications has been enabled by a device called a MODEM, which stands for MOdulator/DEModulator. In simple terms, this device takes a data bit, turns it into a tone (the modulation part), sends that tone out over the phone lines, and then is heard by a modem at the other end, which then turns the tone back into a a bit (demodulation) and sends it on its way to the device on the other end.

This was incredible technology when it hit the streets, and 300 bps modems were all the rage. Even back in the late 1970s, anyone who had a 2400 bps modem was hot stuff. But enter the 1990s and the transition is from the 14.4 Kbps modems to the newer 28.8 Kbps market leader. With some proprietary designs, these devices are operating in the 32 Kbps range, but they have just about reached their limit and the jump to the digital domain is your next obvious choice.

But that jump occurred years ago, when 56 Kbps through 1.544 Mbps transports were made available, very expensive, but fully digital with all of the benefits and performance. These technologies are best suited to data applications, primarily leased lines, exclusive of the integration of voice services down the wire.

Now before anyone yells, there have been a number of proprietary products on the market that have done an excellent job of integrating voice and data in these private networks, and under the right circumstances, I wholly recommend those solutions. But today these solutions are increasingly being based on ATM, the broadband ISDN service, and have moved away from the proprietary transport technologies they used in the past.

Another quick point, many modem manufacturers will tell you that they can transport data at rates much higher than the rated modem speed, through built-in hardware compression they make available. This can yield transfer rates of 56–112 Kbps on normal analog lines. On paper this sounds great, but here is the simple truth. If the data is compressed already, this compression isn't going to do anything. Some formats of data, especially graphics files, are compressed as far as they can go, so the hardware compression in a modem isn't going to help, except for the marketing numbers. With compression, I can get my 14.4 Kbps modem to transfer data at a rate of half a megabit per minute, pretty impressive. But the file is compressed at a 20–1 ratio, so any datacommunications device is going to look good transferring a file like this.

When you are doing your performance comparisons, put compression in its place. It counts for many things, like transferring files across the internet and all of the network traffic data. But while 112 Kbps sounds great from an analog modem, ISDN TAs are hitting 512 Kbps for BRI lines, and that sounds even better to me. So recognize the game everyone is playing with these numbers and judge the specifications accordingly.

ISDN and Datacommunications

There are a number of pieces written and marketing materials running around describing ISDN modems. Problem is, they just aren't that kind of a device. An ISDN TA (terminal adapter) is working in the digital realm all of the time. The information received by the TA is digital in form, and it remains in that form to the other end of the line, when the TA is operating on a data call. If the TA is processing a voice call, for a person talking on the line, or to transmit a fax or analog modem signal, then the TA converts the sound to a digital format, and sends that through the network as a voice data packet, which can be converted at the other end of the call by the proper device.

There is a new breed of ISDN TAs that has embedded modems to give you the most bang for your card and computer resource space, and all at a good hit to your wallet. Manufacturers are delivering integrated ISDN TAs with fax modems and voice services, all in one device, often one card. These carry the modem title, and deserve it because they are multifunction cards that may include a true modem, fax, voice, voice mail, ISDN data, and packet services.

While the ISDN TA transmits everything at a 64 Kbps rate, the V.110 and V.120 protocols provide a mechanism to move data through the circuit at a slower rate, to accommodate the connection requirements outside the TA. If you remember from the standards section, V.120 supports 57.2 Kbps and V.110 supports 19.2 Kbps data rates.

> I/O — Input/Output bus that carries all of the data from your peripherals to the CPU and memory.

The ISDN TA will receive data in one of four ways. The first is through a direct connection to the I/O backplane of the computer. In this connection method, the TA will receive data faster than it can process, so there is no need to worry about the computer slowing up the ISDN line. The second method is through a synchronous interface, where again the TA receives data at the rate that it can transmit it, so you gain the full 128 Kbps rate on a BRI circuit. For our third method, the network connection, the TA can be connected to the network and receive data, again at a speed much greater than its transfer rate, so it will operate at full speed.

The serial port connection operates under some limitations, as it was explained in Chapter 3, the terminal adapter section. The chip sets that handle these transfers have been slowly upgrading in speed from 300 bps many years ago, to a top speed in the 400 Kbps range. But the most popular and affordable products in the United States operate at a top speed of 115.2 Kbps, and this a bottleneck to your ISDN TA throughput. It's actually a little worse than that — the 115.2 Kbps rate includes overhead data, making your real transfer rate 80 percent of that, or 92.16 Kbps. Your modems are subject to the same issue, so a 28.8 Kbps unit really only gets 23.04 Kbps throughput, on a great day.

Connection Types

> **Same-to-Same** — I use this term to mean that the same equipment requiring the ISDN link are operating at both ends, like a LAN-to-LAN connection.

What I want to cover here are the kinds of connections you are likely to make using ISDN and some of the basic issues. ISDN TAs are being used in same-to-same applications, where each end has been set up to operate in the most effective way possible. Your best performance will happen in these situations, because you control most of the issues involved, and you can optimize the equipment to each other, often by standardizing on the same hardware everywhere.

These same-to-same applications can be a single computer to another single computer, or a LAN-to-LAN application. I exclude from this definition the single computer to LAN, because this may involve a mix of hardware and software that makes the connection process more difficult.

The second class of connections I call same-to-different (all right, not real original, but certainly clear in meaning). These include your connecting to an Internet access provider or your local on-line service, or a friend with ISDN. In each case, your ISDN TA is likely to be different than the one you're connecting to, and in the case of the on-line services providers, you are going from a BRI line to a PRI access environment.

And with these connections your access to these other sites with either a single B channel at V.120, V.110, 56 or 64 Kbps, or in a BONDED or MPPP dual B-channel call at 2X speed, or you will use a PRI circuit, with anything from the

56 Kbps to a full 1.92 Mbps. For the last access type, you can connect on any B or D channel using the X.25 packet services, giving you from 16 Kbps to nX64 Kbps transfer rates, and a many-sites to one capability.

Connection Components

In Figure 8.1 each of the components of the data access picture is illustrated. Again we are not looking at anything real different from the requirements of voice. ISDN was designed to enable these services to coexist on the same wire, and using them is simply a matter of accessing the service you want through the correct port or type of TA.

■ Internal Adapters

There are internal adapters that will provide the highest possible throughput for an individual workstation because of the connection from the internal backplane, moving data at megabyte transfer rates, which can keep the ISDN pipe full. There are some areas of concern that must be explored prior to using this version of hardware.

As you see in Figure 8.1, the credit card processor is using an internal TA, which needs an internal NT1. Another requirement is a software driver for your operating system. This is your caution flag. If the card does not have a driver to support it, or if the driver only partially implements the functionality of the card, then you may have something that works at less than full speed or doesn't work at all.

IRQ — Interrupt Request Line, aids in establishing the card that can use the I/O bus.

Another critical concern, mostly with the PC platform, is the availability of I/O slots. The limiting factor, beyond having a physical slot open, is the available I/O configuration parameters. The computer uses three main addresses to uniquely identify an adapter inside your computer. The first is an IRQ, which is limited to a small set of values, many of which are taken by add-on components, like your COM ports, line printer ports, sound cards, net-

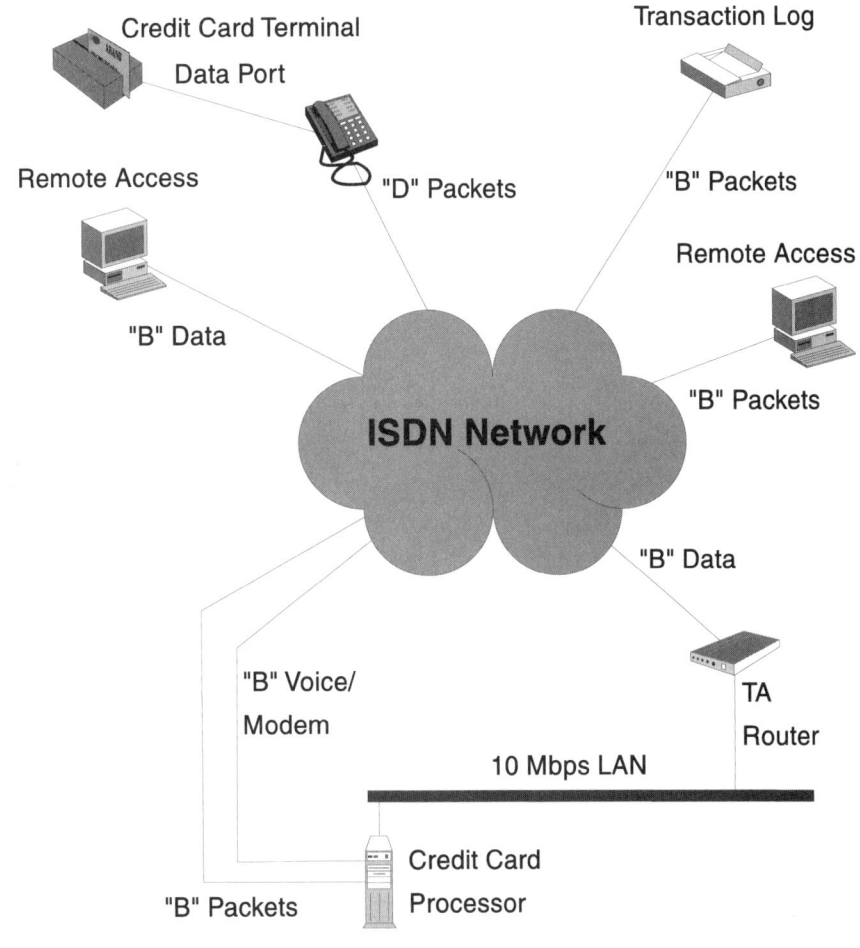

FIGURE 8.1 POSSIBLE ISDN DATA CONNECTIONS.

DMA — Direct Memory Access, this allows an I/O card to move information to the memory of the computer without burdening the CPU.

work card, disk and CD-ROM interfaces, video card, and other internal functions. TABLE 8.1 shows the typical values and assignments for these values. The second value is the Port address, where few conflicts occur; there is a lot more room in this portion of the computer. The final parameter, the DMA address, is used by select devices to access memory outside the control of the CPU, also a limited resource, but readily available today.

It is possible for these parameter issues to consume many hours of your time trying to find a working combination. The tools to identify the correct configurations are not readily available on the earlier PC platforms, but are improved with the new Windows 95 and Windows NT 4.0 operating systems. The PCI I/O bus, where the cards configure themselves with assistance from the CPU, and the Plug and Play technology emerging for the PCs and ever present on the Macintosh platform, assists in letting the hardware solve its own problems, letting you focus on your job, using the technology!

TABLE 8.1 Typical PC IRQ assignments.

IRQ	Usual Functions
1	Keyboard
2	Cascaded Interrupts or available
3	COM port
4	COM port
5	Sound Card, LPT2, Mouse
6	Floppy Controller
7	LPT1, Sound Card
8	
9	Cascaded Interrupts or available
10	Network Cards, SCSI adapters
11	Network Cards, SCSI adapters
12	Network Cards, SCSI adapters
13	
14	Disk Controller
15	Second Disk Controller EIDE

■ External Serial Port Adapters

These devices will connect through the serial port of your computer. In this connection mode, you are operating through a UART chip and dictates the maximum speeds you can transmit and receive data at through this connection type. Many older computers, as recent as 1992, may be equipped with UARTs that are limited to 38.8 Kbps or less capabilities. With the development of the 16550 UART, these speeds jumped to 115.2 Kbps, and proprietary designs support speeds of over 400 Kbps.

> A new serial bus technology is coming to town, the USB, Universal Serial Bus. This bus will have speeds in excess of 400 Kbps, and relieves this bottleneck for you. But to use it you will first have to wait till it is incorporated into new device cards; second, you retrofit your existing computer; and third, you retrofit your existing peripherals. Word of caution, the ability to retrofit peripherals may be nonexistent.
>
> This technology will be great when it arrives. It is highly recommended for new computers and peripherals, because it allows you to put multiple devices on a single port, solving the limited resource issues of PCs. But don't wait to implement ISDN because of this. The current interface options may take extra effort, but it's worth it to take full advantage of ISDN's bandwidth.

The TA is also limited by the UART chip for asynchronous connections, so the maximum transfer rate you will get is 115.2 Kbps less the start and stop bits. But if you are using a synchronous connection, then you are able to use the full 128 Kbps capacity, from both the computer side and the TA side of the interface.

Included in the external TA category are the ISDN phones with data ports, through which you can access packet-switched data services or establish a direct connection with another ISDN phone that has a data device attached. This is a highly functional configuration for applications like remote printing of security events, credit card transactions, etcetera.

■ Network Adapters

This is by far the best use of an ISDN TA, because placement of the TA on a LAN provides access for all the computers on the network, not just your workstation. Although it is possible to use a computer as a router, this can be an ineffective use of resources, especially if the computer goes down, your router will go with it.

The connection complexity changes dramatically, because the TA is talking to a network. Networks use a variety of communications protocols: TCP/IP (the Internet), SPX/IPX (Novell), NetBEUI (Microsoft), and others. The entire topic of network protocols fills many books, several are listed in Appendix C if you want to learn more about these. This communication occurs on an Ethernet port, currently at 10 Mbps speed. It requires that the terminal adapter be configured with a network address, and that it has the correct protocols to talk with your servers and other networking devices. Lastly, your computers must have a network link, with the same protocols, in order to use this device.

In addition to basic networking protocols, there are protocols for remote access services, with PPP and SLIP being the most dominant. PPP is becoming the standard for remote network access. You then add security authentication protocols like PAP and CHAP, plus other security features to validate the person trying to connect to your network, and you can see the complications. But the rewards of your entire network being able to share the ISDN link make this option worthwhile.

THE APPLICATIONS

The Store

Perhaps one of the best examples of using ISDN in its simplest form is the small store that wants to process credit card transactions, and do a few other things: for instance, handle phone calls, connect their computer to various on-line resources, and work at home. This is a great multipurpose application, because it touches on all of the ISDN services.

Figure 8.2 shows the connections of equipment on the site and what function each connection line is providing. From this example you see that each line has been maximized for use, with a single BRI providing all of the functionality. This is possible

PRESENTING THE DATA APPLICATIONS

because of the working patterns of the store owner and the ability to do the various activities at different times of the day.

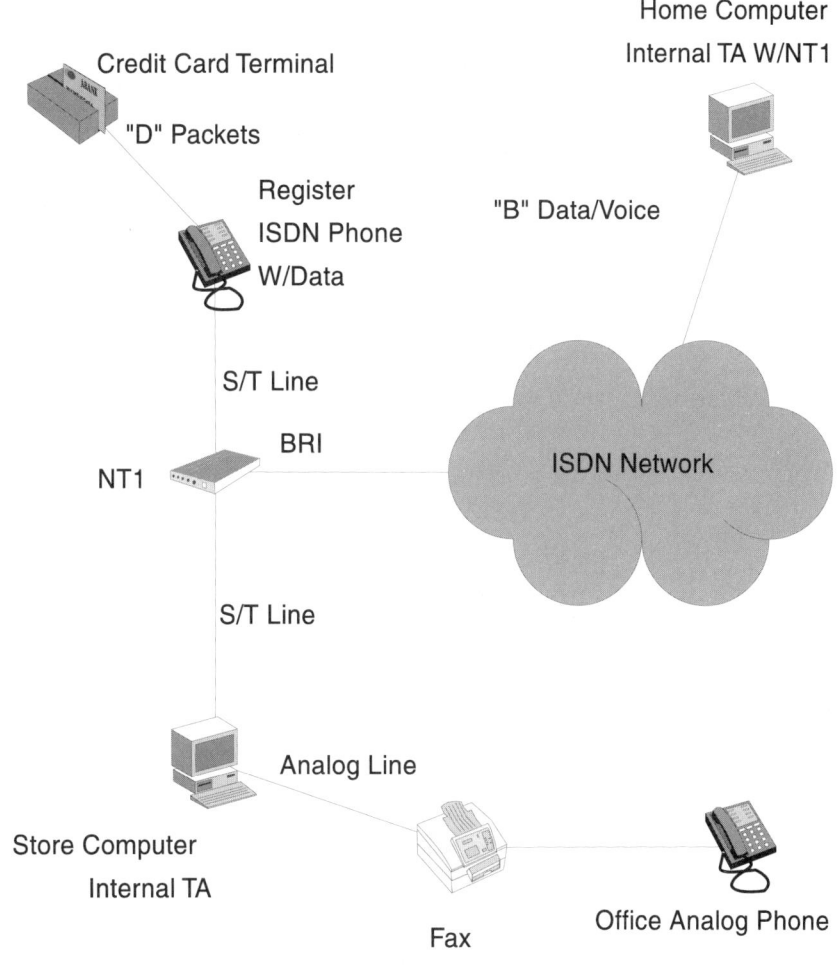

FIGURE 8.2 STORE EQUIPMENT AND LINE CONFIGURATION.

During the day there are two priorities in this store, the authorizations of credit card transactions and the handling of phone calls. For this functionality the D channel packet service will handle the credit card transactions. The credit card service, which also handles check authorizations, has a connection to the public-packet network, and the store has access to that same network.

With a credit card terminal attached to the data port of the phone by the cash register, the transaction and credit card information is entered, the send button then establishes the packets and transmits them over the D channel. The terminal then processes the response a few seconds later.

During the day, the B channels are reserved for the voice services. With two phones installed, one out front and one in the back office, two calls can be going at the same time. With the multiple call appearance features, the phones can handle two calls each, giving the store the chance to serve four people at a time.

To handle the data requirements, a TA is installed in the computer, allowing for B-channel circuit-switched data calls. The owner does three things with this. The first is to connect to suppliers through an EDI application, to transmit purchase order information, receive order confirmations, and shipping notices. The second application is remote access from home; so when the store is closed, the owner doesn't have to stay at the store to update records and manage the business. For the third use, the owner, like many other people these days, wants to surf the Internet. The ISDN line at the office provides that opportunity. In addition, the products being sold in the store just might sell well in a mail order environment, so there is a possibility that the store may go on-line with a Web site of its own.

> **EDI —
> Electronic Data
> Interchange.**

Setting up this application is very straightforward. To perform the data applications, a data port is needed on the phone to allow for the packet-switched services. For the computer access, a TA for the PC, external or internal, is needed with an analog port to support the fax and modem. And by using an analog phone, there will only be two devices connected to the NT1. The last piece of the puzzle is ordering telephone services so the phones can act in unison, with call hold, transfer, hunt, and direct dialing services (Centrex would be ideal, but this service usually requires two separate lines as an order minimum). And the last is a protection item, two UPS units. One for the cash register area, keeping the phone up for emergency purposes and allowing the cash regis-

Presenting the Data Applications

ter to be closed out in an orderly manner if the power goes out. The second UPS supports the PC, NT1, and TA, again to provide an orderly shutdown of these devices and provide emergency phone services during a blackout.

To enable the system to system access, a BRI line is installed at home, with voice and data services on both B channels. You can refer back to Figure 7.7 for ideas on how the home line can be used. Then, in the computers at both ends you will use some remote access software, like PC Anywhere from Symantec, to establish and manage your remote connections. If you are using a product like Windows NT from Microsoft, the remote access software is built in. Use Windows NT at work, with Windows 95 at home; this makes for a great PC combination these days.

For the on-line services you can take your pick. As long as you use the proper connection configurations, you will be able to access all the major providers through ISDN at home and the office. And Internet access is just as easy. Many Internet providers have dialing packages to support setting your PC up with the proper protocols for their network, frequently V.120 or PPP, and do all of the connection work for you. Even the Windows NT environment can be easy to use. Dialers are coming to market to work with this system now, and the next generation of NT will have enhanced auto-dial capabilities to connect to many services.

Remote System Access.

In the previous example you saw some remote access capability. But what if your needs are greater than a simple system to system setup? This application story will help you understand how to accomplish this task and what you will have to account for in making this solution real for you. As a tickler of the value of this service, take a look at the application story from GTE and Tektronix.

We'll look at another application story in a while, but let's see how you can do this same application for yourself. To begin,

■ THE ISDN CONSULTANT ■

In Oregon there are financial incentives from the state to implement telecommuting solutions, with a 35 percent tax credit against the cost of the installation and equipment to make this application possible. With that incentive in hand, Tektronix went to work with GTE to determine the best way to get about 150 employees to work from home.

Engineers and I/S support people are participating in the program, using Ascend HX multiplexors at home and Ascend MAX units at the central site. This gives BRI as the remote link at home connected to a PRI link at the central site. The initial implementations have been data only, but the plan includes implementing TAs with voice capabilities to expand the flexibility of the solution.

This solution required about $1,500 per home in start-up cost, and the coordination of multiple local phone companies. But the application has been up since the pilot in 1995. The initial pilot brought 15 employees on-line over a three-month period. Since everything was worked out successfully in the pilot, and the investment proved itself, the program has been expanded.

In this project there is a reliance by Tektronix on the GTE team to oversee the installation of all sites, freeing Tektronix to manage the productivity of the program, and letting GTE assume the details that can bog down a project like this when you are still in the learning curve of the technology and social changes that this technology brings about.

Figure 8.3 shows all of the basic components of this solution. The key component here is the central site device that facilitates a number of remote access users simultaneously. Instead of putting a series of BRI devices at your central site, when you get past about six remote site users, you should look into a PRI solution instead.

You'll want to do this because the cost of TA units that support PRI are a fraction of the cost of 12 BRI TAs to support 12 users at 128 Kbps. And with PRI comes simplified line management and flexibility. You could easily mix single and double channel access down the PRI pipe. And with the proprietary bonding that the units will do in a private network installation, you can bump your bandwidth up by a factor of four with some of the solutions on the market.

Another technique to enhance the benefit from this solution is to consider Centrex services for the lines in this application. If you remember the discussion in Chapter 7 about Centrex, you

PRESENTING THE DATA APPLICATIONS

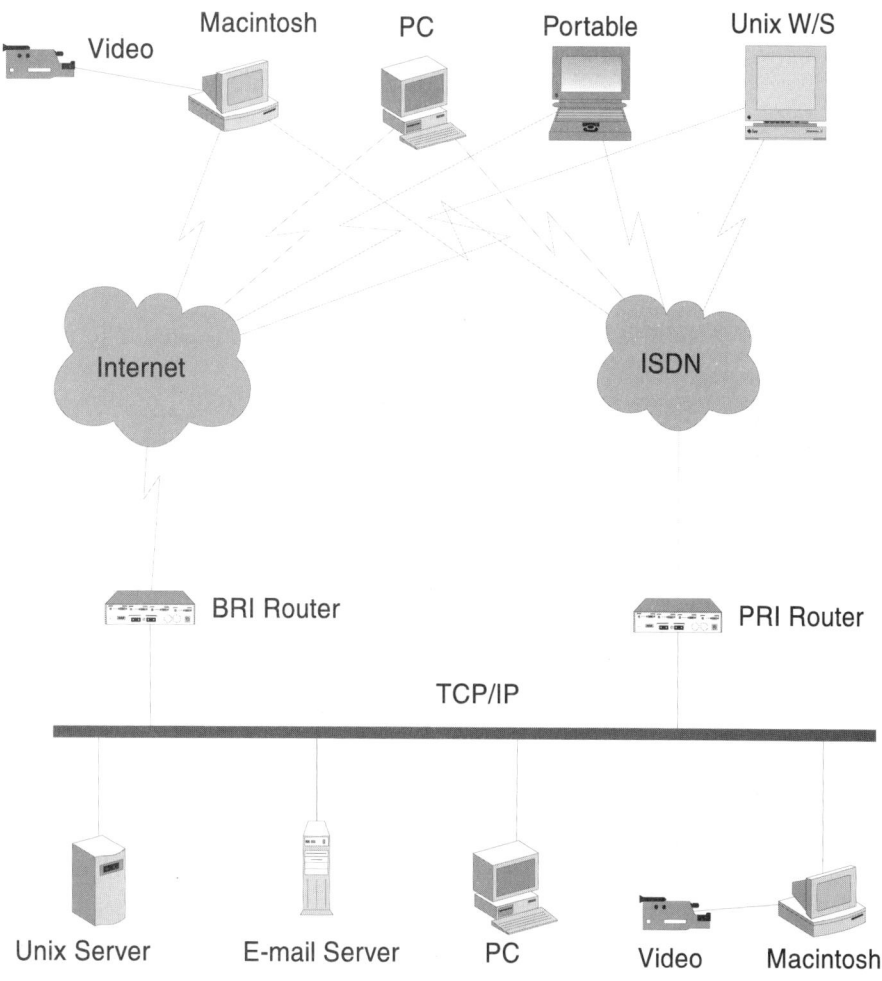

FIGURE 8.3 REMOTE ACCESS APPLICATION, WITH INTERNET ACCESS AND VIDEOCONFERENCING.

know that the line is a flat rate for calls between phones in that *office*. In this case, the idea of an office is extended to the off-site locations. Centrex will fix your monthly overhead, allowing you to provide unlimited access to your network at this constant cost. The balancing act you have to pursue is the mileage when your Centrex sites go beyond the coverage areas of your CO. At some point the mileage charges will exceed the expected usage or leased line costs; thus there is a need to determine that point and choose your solution accordingly.

As you can see in Figure 8.3, the central site will use a router device connected to either PRI or BRI lines. Some of the routers in the market will provide for multiple BRI connections, giving you an option for situations where you need three or four lines and not a full PRI. What do you need to watch out for?

One of your primary concerns should be with the security of this application. A number of routers will provide for a dial-back capability, and this is a limited but effective solution. On top of this network you need to lay down other security measures. For example, the simple act of logging into the network requires passwords and authentication provided by your network operating system. These measures will chase away many intruders, but not the serious cyber-criminals that may find your site today. For these folks you need to add a layer of security both in the logon process and in the transmission of data.

Security can be configured through public key encryption schemes, where you and the person you are communicating with have a public key that you exchange with each other to allow the exchange of information between your systems. Through a bit of math and technology, these public keys are combined with a private key you have, and this creates a secured packet that is only readable by the partner in the transaction. While all security schemes can be broken, the effort required for the emerging security measures will push away all but the most serious intruders and the curiosity seekers that do this for the challenge.

It's always a concern to add support for remote systems. Any time you add devices to your network, you can create problems if you don't put certain safeguards in place. As unfortunate as it is, there are computer viruses that can infect your systems, and quite often they come in through the innocent actions of an employee, who may have picked up a file from the on-line services, which then tags along with an office file into your network. You will need to plan on putting virus scanning software at each remote computer and screening all incoming files to safeguard your own site.

Another matter that may be overlooked is the legal issue and liability you may incur having employees officially work at home. By doing this, you are extending your property and responsibilities to their home. And in their home you have little control over the working conditions and safety of those people. Should anything happen while they are on the job at home, you will be held responsible. This could be from a work injury related to the activity that they are involved in, or it could be from a fall tripping over a toy that was left in their path while they rushed to answer the phone. Either way you may have to cover that employee for all of the damages and lost time. So take some time with the human resources and legal folks to establish good policies and safeguards in this area.

LAN to LAN Services.

The LAN to LAN services are a bit different, because you are working with multiple systems on each end. In this application, the devices will be routers on either site of the ISDN link, and the routers will manage all of the traffic for you. Some special considerations need to be made when you are connecting networks.

Networking protocols are designed to work on a local or wide-area basis. The local area network protocols such as Appletalk and NetBEUI can only be bridged at this time. Microsoft is reported to be working on a routing capability for future versions of the NetBEUI protocol. The TCP/IP and SPX/IPX are routable protocols. All of these protocols were designed with the expectation that the links between networks, whether local or remote, would be always available. With this expectation, the protocols will issue packets to poll the network for activity and information. With SPX/IPX this activity happens very frequently and contributes to the total traffic level on the network. Whenever these packets are placed on the network, the bridges and routers will attempt to transmit the packets to the remote locations.

ISDN routers and bridges, on the other hand, have been designed to drop the link whenever possible. This feature is meant to minimize your connection cost by transmitting only packets that need to be sent. The question now becomes one of discriminating between the persistent traffic that doesn't need to be transmitted and the real thing.

What many of the ISDN LAN bridges/routers (TA) have the ability to do is fake the local network into thinking that everybody on the remote end is alive and well. This concept allows the TA to issue packets that let the network devices, computers, services, etcetera, operate as if the remote network is present. Then when packets arrive that need to be transmitted, the TA brings the link back up, within seconds, and the link is maintained until traffic ceases and the link remains idle for a defined period of time. Other techniques exist in the world of routers to prevent unnecessary traffic from being sent out over these links. If you are interested in knowing more about this, take a look at some of the networking companies' Internet sites, like Cisco, Ascend, and ACC, where a number of excellent white papers and technology briefs are available.

One last use of this dial-up on demand principle is a leased line backup application. With this, if your leased line goes down, you can switch to an ISDN device that will dial your remote site and reestablish the link. This typically happens with some operator intervention and may force you to restart the activities that were active when the lined failed. But this technique gives you a fast recovery time while your networking staff and the leased line provider figure out what went wrong.

■ What's Next

Now that your voice and data is coming together, let's look into the world of videoconferencing. The only real changes to the game will be the devices we use to perform the application, not the ISDN interconnection devices. With that in mind, you can see that this technology is truly flexible and easy to extend to other applications; just get the basic interconnections going and you are in great shape. So on to the last chapter and have fun.

CHAPTER 9
Presenting the Video Applications

As they say, a picture is worth a thousand words. But a voice and moving picture are priceless in my opinion. How much more can you communicate when you can see the person on the other end, watch how they react, see what they are trying to explain by way of body gestures, the material in hand, or even the act of demonstration. Nothing, except being there, beats seeing and hearing someone across the line.

Videoconferencing has been around for many years, but not until recently has it gained market and popularity. What is driving this is the significant gains in compression technology, bringing near broadcast quality to transmission speeds under the PRI level, and the emergence of desktop conference stations, for under $2,000 with acceptable transmission characteristics across a BRI line. So let's take a look at the technology and then dive into the simple task of implementing this application.

■ Videoconferencing Primer

In a sense, there isn't much to videoconferencing systems; they can be as simple as the other applications you have seen so far. But what has made them simple today is the advent of standards, improved chip technologies, and the availability of high-speed lines like ISDN.

In the early days of videoconferencing, you had to pay $50,000 or more to get a conference room system, and you had to have a T-1 circuit to get good full motion video. This made the technology much too expensive for everyone but the largest of organizations, because the monthly recurring costs for the lines far exceeded the occasional benefit of the videoconference. Yet the industry knew that these problems could be solved, and solve them they did.

The first effort was to create enhanced compression technologies. This process creates the ability to get a better picture out of the existing bandwidth and opens the door to a lower quality picture on a much slower line. At the time, these compression technologies were proprietary, so you had to have the same units on each end, making your solution an internal one, and just maybe, you could use them with key outside trading partners if they agreed to buy the same products you did.

■ Standards

The industry recognized that proprietary connectivity didn't make any sense. Everyone was looking for interoperability, and the H.320 videoconferencing standard emerged. Through this standard, you can take any complying product, dial it up through an ISDN link, and carry on a videoconference. This opened the door for the manufacturers to work with the computer folks, to begin the marriage of the PC with the video technology. As prices began to drop, conference room systems started appearing in the $20,000 range several years ago, and desktop computer units were coming in at about $5,000.

PRESENTING THE VIDEO APPLICATIONS

With ISDN making speeds from 64 Kbps up through 1.984 Mbps a reality and compression technologies performing even better, the door opened for industry-wide compatibility. For the first time, users could call anyone with a matching system standard, all for the price of a nX64 Kbps call. This level of affordability coupled with quality seeded the demand for the technology.

But the standards continue, as the newer T.120 standard deals with the issues of the multimedia conferencing experience. This is a recently approved standard, attempting to address additional issues the H.320 standard does not. In addition to that, you have the MPEG-2 standards for video compression in the 2–15 Mbps transmission rate range. The MPEG-2 standard is being hailed as the tool to deliver video on demand across our broadband ISDN networks.

All these standards combine to provide a compatibility between videoconferencing solutions, allowing you to purchase equipment from the desktop to the boardroom, and let the users communicate whenever they need to.

■ *The Basic Components*

At the heart of the videoconference systems are the CODEC subsystems. These COmpression DECompression subsystems, CODECs, provide the translation of video and audio signals into the multiplexed format that can be transmitted by the ISDN line. These units take your video and audio signals and move them through their own CODECs, where the signals are then combined and passed on to the ISDN line. The control functions, such as dialing, remote camera controls, etcetera are also added to the data stream that is being sent, providing complete end-to-end conferencing functions.

As you go beyond the desktop systems, where a system can cost as little as $250 and give you a few frames per second over an analog line, to full room systems costing $20,000 or more that

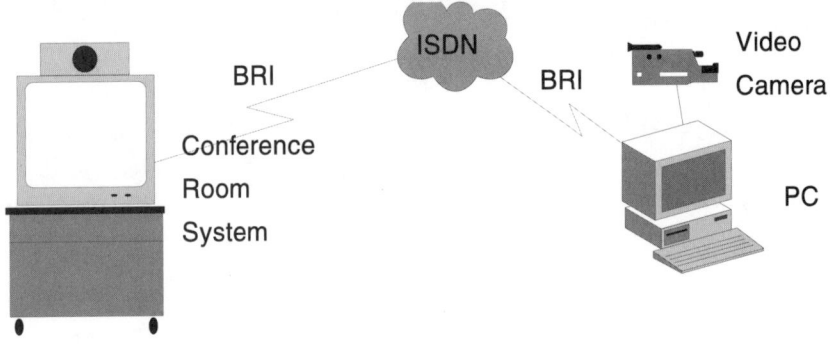

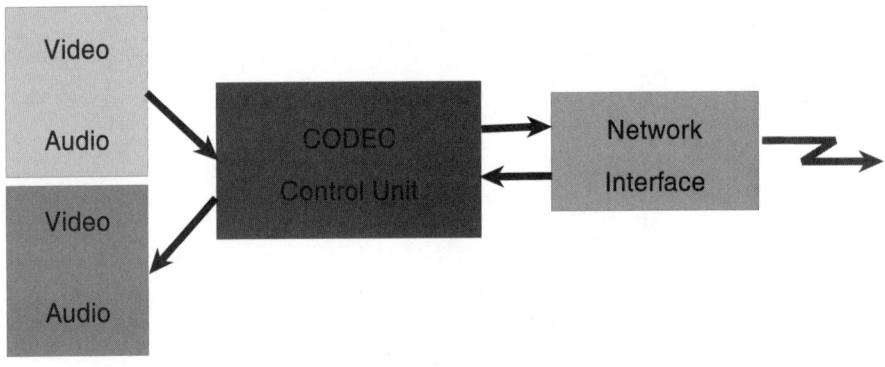

FIGURE 9.1 BASIC VIDEOCONFERENCING ELEMENTS AND CONFIGURATION.

give you full motion video over 384 Kbps to 1.984 Mbps calls. In each of these systems will be:

- Video Camera
- Microphone(s)
- Camera Controls
- CODEC Unit with Network Interface
- Monitor and Speaker System

These components are typically not mixed and matched, but instead are purchased as a complete solution from the vendor for conference room systems (see Figure 9.1). For the conference room, you want to look for certain qualities about the camera and sound subsystems. For example, if you have a camera subsystem that has a follow-the-voice technology, then you can count on the camera following you as you speak and pace around the room. But if you only have a manual camera control, then you're going to have to stay in place or have someone play follow the leader.

On the voice side, you want a microphone that has good pick-up qualities from a number of feet away. If the microphone only catches the conversations in the first five feet, your conference will be interrupted as people get up to come to the microphone, or you will have to pass the microphone to everyone, both a hindrance to a good videoconference.

For larger conference situations, you want equipment that can distribute the video signal to multiple monitors and speakers, letting you do a miniature broadcast in your facility or to enhance the viewing in a larger room.

For desktop videoconferencing, your computer's monitor and sound-card speaker system provide the missing components to operate as an H.320 compatible videoconference site. Your conference kit provides the video capture and CODEC function, an ISDN interface, and the software.

There are special control units called multipoint, which as the name implies, distributes videoconferencing sessions to multiple locations at the same time. Through these units, some with up to 32 channels, you can link multiple sites, each at different speeds, up to a full E-1 rate. These systems allow you to develop a conferencing strategy that brings people into regional locations or public videoconferencing sites, and tie them altogether through your facility or through a public service that can provide this service. These high end multipoint servers can cost as much as $250,000 for a 32-site conference unit, not including the cost of the actual videoconference equipment at each site.

Those are the basics for the technology and standards. Your key to success will be ensuring that all of the equipment you purchase meets the same standards, and the standards of the sites you want to connect with. Make your decision for level of product you want based on a compromise between product cost and quality of the video. Always stretch as close to full motion video as you can afford.

Even the best of equipment will do you no good if you don't have a good working environment to conduct your videoconference. This means that you must provide adequate lighting to illuminate your subjects and other objects you will want on camera. The acoustical qualities of the room also effect the overall experience. Hard walls and floors will cause the sound to bounce, contributing to feedback or an unnatural sound. Carpet and acoustical treatments for the walls and ceilings can help to dampen the sound.

But enough of the background, let's look at the applications themselves, and see how simple they really are.

■ The Applications

Conference Room Systems

This is where videoconferencing got its start, and where the technology remains the most exciting. As these systems dropped in price, the financial return to own one came within reason for many people and organizations.

These systems can be used for the simple business functions of holding weekly sales meetings with managers from around the country, or they can be used to deliver distance learning over large geographic areas. A competing transport technology to ISDN in this arena is satellite broadcasting across large areas, and local Cable TV systems. Several problems are resolved through ISDN that may impact upon you in these other two technologies.

Satellite broadcasts are subject to delays, based on the time it takes to transmit the signal 22,500 miles up to a satellite, relay it back down that distance, and then return to you with the image of the person reacting to your picture or words. This delay introduces a half-second delay for the signal to travel to the remote site, plus the delay to process the signal at each end, often giving a 2 to 5 second delay, which can result in a lot of people walking over each other's speech. Over land lines, these delays all but disappear, leading many companies back to transmitting their video and audio signals over ISDN or other broadband ISDN solutions.

> The power of broadband ISDN was clearly demonstrated a few years back, by Steven Spielberg, when he was completing work on Jurassic Park. At that time he was in Poland to film *Schindler's List*, while the ILM studios were doing final work on the special effects for Jurrasic Park. By setting up a 45 Mbps line between the two locations, Mr. Spielberg was able to work with the special-effects editors, reviewing and directing the modification of the film in realtime and cinema screen quality.
>
> With the enhancements to compression, recent demonstrations of 768 Kbps and 1.536 Mbps video transmissions appear as good as most broadcast quality images you receive today.

On the other hand, the Cable TV solution can work nicely when you are trying to broadcast to many sites in a moderately interactive arrangement. Many early Cable TV interactive sessions were accomplished via the dedication of video channels, where each channel provided a one-way broadcast to the other site. You can see the limitation in the strategy: Only two sites get to interact, and to watch both sites you need two TVs.

In the application example for this technology, rather than understanding a corporate conferencing environment, or a traditional distance learning application, let's understand what happens when you combine several environments into an interactive learning adventure. In Figure 9.2, I have shown a skeleton of a distance learning session that involves a central site with a live whiteboard, an audience camera/still camera, a monitor and audience camera setup, and a multipoint capability to handle five remote sites for this session. In the first remote location we have a camera transmitting over a wireless link from a diver in the Pacific Ocean, back to a land site, which is connected by an ISDN link to the central site. Three of the remote sites are 128 Kbps PC desktop videoconferencing systems, while the fourth site is a monitor system with a separate audio call to provide feedback.

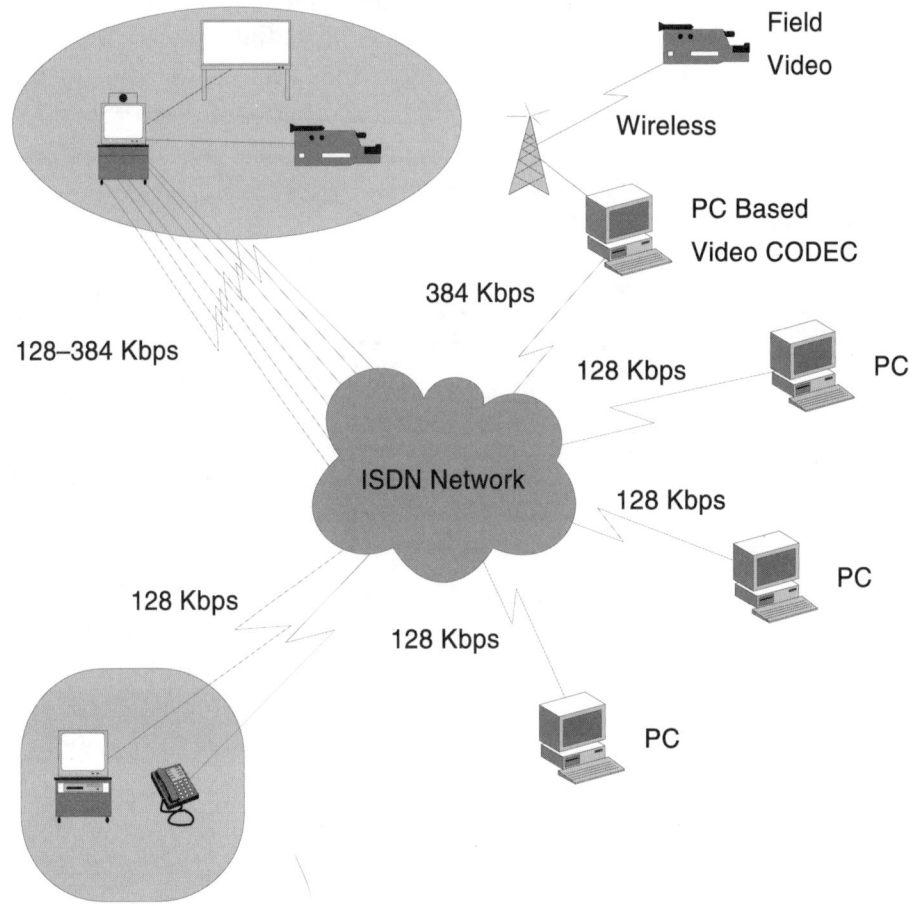

FIGURE 9.2 DISTANCE LEARNING VIDEOCONFERENCE WITH WIRELESS FIELD FEED.

Let's look at some of the basics in this configuration. The central site has a number of connections and devices to make it work. The first key device is the multipoint control unit to allow for transmission to more than one location. In this arrangement, each remote location will have its own line coming into the multipoint controller. This device will broadcast the video and audio

PRESENTING THE VIDEO APPLICATIONS

packets across all of the lines, each at the appropriate speed. Add to this the still camera in the room to transmit slides and other presentation materials. This video source will be selectable through the controller unit at the site. The third item is the Whiteboard, which can transmit the information written on it in the form of screen images, allowing the remote audiences to see this information through compatible software on their computers. A fourth source of information will come from the field feed, again selectable from the conference controller station.

At each of the remote PC sites, a desktop videoconferencing system, operating at 128 Kbps will provide a good near-motion video experience, with perfect audio to listen to the conference. The PC system will allow the users to participate in the conference on a two-way basis. Each PC has a limited field of view camera and can display the conference on the desktop monitor. Segments of the conference can be captured to disk, allowing the participants the opportunity to archive key pictures and diagrams at their discretion.

The remote site with a video cart uses a video only monitor unit, and a telephone for the audio to ask questions. The video cart has its own CODEC and is receiving the broadcast at 128 Kbps.

The field site actually operates on a desktop system, at the higher 384 Kbps to give a good feed for the video being received from the wireless link. The video is taken underwater and fed to a support ship, which then transmits the video through wireless to the PC station onshore, which then feeds the video signal onto the link to the central site, where it can be monitored and included when desired. This system also allows for the people in the classroom to ask questions or participate in decisions about the activities of the diver, in real time.

Desktop to Desktop Solutions.

Desktop video is a new revolution. The affordability of workstation solutions at under a $1,000 makes it a very attractive

> While this may seem like a fantasy, variations of this activity are happening now, in places like Dana Point, California. As an even better twist to the application, the video feed is sent out over the Internet through the Global Schoolhouse project.

> But don't let your imagination stop there. I challenge you to this idea in distance learning. Go out to your local water districts and see what kind of conservation programs they have. Many I know of or have heard about operate sensitive ecological preserves, where they bring in school children to tour and educate. The sad part of this situation is that they can only describe ecosystems in other parts of the country or world.
>
> So let's do something about it! Get them to install camera systems, on a wired or wireless basis, around their ecopreserves. With remote control facilities, they can capture things in their natural state, such as migratory birds and other visitors. Then bring all of the camera feeds back to a video control booth, where the entire project can be monitored. From there connect to the world through some ISDN links, scheduling remote visits from schools everywhere. Even better yet, if you use this video feed with the Streamworks technology from Xing, you can put it up on the Internet, where anyone can view it with the viewer Xing provides.
>
> What kind of appreciation for nature and for the different climates of our country, and other parts of the world for that matter, do you think this would make. I believe that you can entice people through the see-it technology, and once enticed, they will be more likely to go touch it, taking us back to the High Tech — High Touch paradigm.

alternative to travel. When you combine this with the low cost of ISDN as a transport medium, you get the right ingredients for better collaboration on a project and a better result from a project team, because many of the details and issues will have been discovered through this visual, keep-in-touch technology.

Travel has always been a motivator for people to justify videoconferencing, but I have talked with companies that see this as a necessary tool even for small campus sites where people are located across several blocks, and it is more productive to strike up a videoconference than it is to walk across the courtyard. This is especially true when the courtyard is in Minneapolis in the middle of the winter, like 30 degrees and snowing.

PRESENTING THE VIDEO APPLICATIONS

Let's run with this idea and see how we could configure a creative solution, built upon the Centrex ISDN services. In this situation, Figure 9.3, the users will have an NT1 with two S/T ports to support their ISDN telephones and desktop videoconference interface. With that in hand, each user has a dedicated BRI line to give them 128 Kbps videoconferencing capabilities, at the exclusion of their phone calls. Don't worry, calls are either being routed over to their backup or out to the voice mail service, so this isn't a problem.

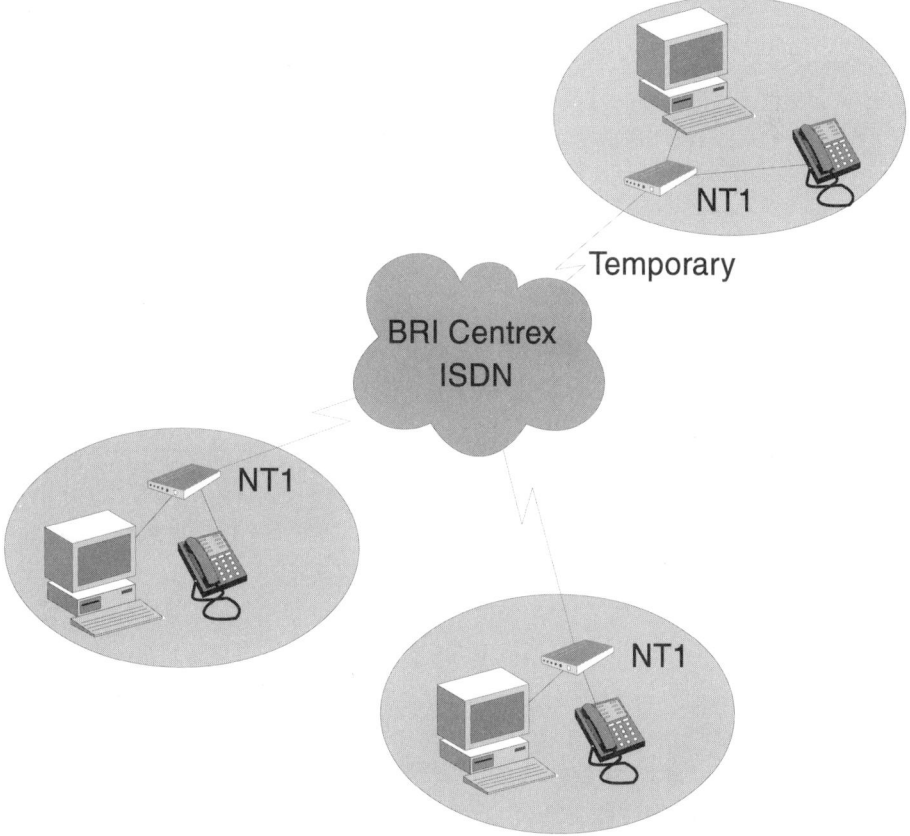

FIGURE 9.3 MIDWINTER CENTREX VIDEOCONFERENCE APPLICATION, THREE SITES, TWO COMPANIES.

When it is time to initiate a videoconference, the user contacts the other person and confirms the start of the conference.

Then both parties hang up and set their videoconference software into action. The two sites are quickly connected, about 300 milliseconds again, and they carry on the visual conference as planned. When everything is done, they drop off and begin taking voice calls, and are ready for any other ISDN-based activity they want to do.

Your cost for the call — FREE! Remember, Centrex traffic between two extensions in your Centrex virtual office carry no usage charges. It's the same as if you had a private phone network to carry your traffic. A system like this will pay for itself the first time you avoid a flu epidemic, because your employees don't have to fight the weather any more than necessary.

But the benefits don't stop there. It turns out that you have a major customer just a few miles down the road, working on a project with them for the next six months, October through March. You want to have weekly project meetings and daily updates on the progress and designs. Just getting out with the team in this winter environment, across the river and out to their site takes about an hour, making the on-site meetings a costly event. By going to the phone company and getting a Centrex line again, this time for a temporary period of six months, you can fix your costs for the conference and phone activities between the two companies. And when it is time to review materials, a quick call puts you into full view to get the work done.

More importantly, with the Whiteboard software, you can actually share design information and documents on-line, letting both sides manipulate the information, bringing faster closure to problems and design changes, most likely leading the project to completing on-time, on-budget, or even better, early and under-budget. I can't think of a better position to be in to win another contract from this company or another.

PRESENTING THE VIDEO APPLICATIONS

> Hudson Foods, a chicken manufacturer located in north west Arkansas has discovered a market opportunity in Poland. It seems that the people in Eastern Europe are partial to dark chicken meat, while consumers in the United States prefer the white. So Hudson headed off to Poland to set up a distribution center. They ship the chicken meat from New Orleans to Poland, and then out to Eastern European countries and Russia.
>
> It became very evident that keeping in touch with this start-up operation was important, and doing so face-to-face was going to be necessary, but the thought of people flying back and forth led the company to look at videoconferencing as an alternative. Working with AT&T, the company entered into an Alpha test of ISDN services to Poland, and as it turns out, had the second ISDN in their headquarters area installed, just behind Tyson Foods.
>
> The videoconferencing system was an AT&T Visitium, with a BRI circuit at both ends. AT&T did the complete line installation management in both countries, coordinating all of the carriers and technology. The end result when AT&T said the line was ready and the equipment installed, Hudson turned on the conferencing system and dialed out. Success on the first call, connected right through. As the first channel went up, everyone was pleased, and when the second channel kicked in and the video smoothed out, the project was declared a success.
>
> Now that the technology and the immediate need has been met, Hudson is looking to leverage this dial-up capability. With the installation of an HP Netserver and Cisco 1003 router, they are going to interconnect their LANs for file transfers and E-mail activities. The use of a Tone Commander NT1 allows the two functions to share the BRI, again maximizing the use of this technology.

Video telephones

What strange creatures these be? For those of you old enough to remember the early sixties, AT&T has been trying to get us to believe that the age of video phones has finally arrived. Actually, they tried almost 30 years ago, then gave up for over 20, and have come back to the market a few times since then with several different models. Most of the recent entries have been limited to an almost still picture capability, across analog lines, until this past year. BT introduced a nice ISDN-based phone that had a roughly 3.5 inch LCD display, including full voice, a limited camera range, and a relatively good picture quality combined with consistent sound.

Implementing ISDN picture phones is a snap. You just order up the line per the manufacturers' specifications, add in the equipment, plug it in, make sure you have a partner on the other end, and you are in business. The downfall, these units don't come with data ports; they are a simple video and voice transmission device, so they are limited in application.

My impression — the desktop video market may eliminate the possibility that video phones in the office or home may ever take off. The fast-paced integration of ISDN with video and computers may lead to public pay-as-you-go computer stations that also have video capabilities, becoming the de facto standard for public video phone functionality.

■ WHAT'S NEXT

The last chapter before the resource appendices will go through the opportunities of mixing all of the applications into a single set of lines. In the three application chapters and throughout the book as the technology was explained, you know that the real power of ISDN is the multiple function capability, and the next chapter is only going to reinforce that position.

CHAPTER 10
Putting It All Together

After spending the last few hours together going over a lot of technical and application information, you begin to get the idea that ISDN is a technology that can do many things, and that it's possible that it is highly underutilized. A piece of information I gathered in the last days of writing this book reported that ISDN in the United States was being used 80 to 90 percent of the time as a data service, while in Europe the ratio is 80 to 90 percent as a voice service. You wonder, what does each side know that the other doesn't?

The commitments of the various carriers to the technology, and more importantly, to the applications they support, drives the initial market acceptance and use of the products. In the United States ISDN has been a very late bloomer, coming into popularity only since the early 1990s, while Europe and Japan have far exceeded even our best hopes for ISDN market penetration in the next few years.

But that tide is changing now. The number of equipment manufacturers, telephone carriers, and some of the new technologies, like PCS, are all tightly woven around ISDN as a key technology with the range of bandwidths that need to be available to service the market. This gives ISDN a bright future and says that we all need to look closer at the way ISDN can be implemented in our organizations today, and recognize the valuable place it will hold for a number of years to come.

■ THE FINAL MIX

Putting it all together, seems that it has already been done many times throughout the book. And in each of these applications you saw that there were some key issues or details to take care of when you were setting up applications to share an ISDN line. Many of the issues are technical in nature, and as you know, this book was intended to give you the depth of technical knowledge to envision the applications and work on the projects with the folks who understand all the nuances and anomalies. I refer you to the ISDN Applications Catalog on the CD-ROM from the NIUF, and to the other technical books on the market to satisfy your burning technical curiosity.

I encourage you to review the various applications here, and in the ISDN Applications Catalog on the CD-ROM. They should give you many ideas on how you can use ISDN to better your business and personal life. And as you do so, recognize that you need the consultation of your staff, your vendors, and the telephone companies to make sure that all of the standards, software and hardware capabilities, and resources are in place to make your application work.

■ WHAT'S NEXT

The appendices that follow will provide you with information on resources on ISDN equipment and software, the telecommunications companies, and other books detailing Internet and on-

PUTTING IT ALL TOGETHER

line resources, and organizations to help you in your ISDN quest. For another 500 plus pages, make sure you visit the CD-ROM to read and use the NIUF Applications Guide, where many of the industry experts have contributed to a highly technical and informative resource to help you through your search for information. On it you will find hundreds of vendors and products not listed in these pages. Appendix D will guide you through the particulars of using the CD-ROM and installing the viewer technology.

Thank you for spending the time with this book, and I hope that it has helped you to expand your understanding of ISDN and your vision for using it to solve the challenges of today. If you have comments about the book, questions you would like answered, please send an email to me at *roblee@pacbell.net*, with *The ISDN Consultant* in the Subject field.

APPENDIX A
Hardware & Software Resources

It all comes to life based on the real products that you purchase to make this technology work. Here you will find a number of resources gathered to guide you to a starting point for building your ISDN application. While the list is long, it is by no means exhaustive. To supplement this I have included a copy of the North American ISDN Users Forum Application Catalog on the CD-ROM. There you will find hundreds of pages of resources and product information to meet your investigative needs.

■ TELEPHONES

TABLE A.1 NT1s.

Vendor	Phone	Internet Address
Adtran	800-788-5408	http://www.adtran.com/
Lucent Technologies	800-432-6600	http://www.att.com/lucent/

TABLE A.1 NT1s. (Continued)

Vendor	Phone	Internet Address
Alpha Telecom, Inc.	205-881-8743	Not Available
IBM	800-426-2255	http://ibm-direct.e-com.ibm.com/us/network/
Nortel	800-667-8358	http://www.nt.com/
Siemens Stromberg-Carlson	407-955-5000	http://www.ssc.siemens.com/
Tone Commander Systems	800-524-0024	Not Available

TABLE A.2 Telephone suppliers.

Vendor	Phone	Internet Address
Lucent Technologies	800-432-6600	http://www.att.com/lucent/
Cartelco Inc.	901-365-7774	Not Available
Fujitsu Business Comm. Systems	800-553-3263	http://www.fujitsu.com:80/FBCS/
Lodestar Technology, Inc.	800-378-6316	Not Available
Motorola ISG		http://www.mot.com/MIMS/ISG/
Nortel	800-667-8358	http://www.nt.com/
SiemensRolm	800-765-6123	http://www.siemensrolm.com/
Siemens Stromberg-Carlson	407-955-5000	http://www.ssc.siemens.com/
Telrad Communications	516-921-8300	Not Available

■ DATA NETWORKS.

TABLE A.3 Data Networks.

Vendor	Phone	Internet Address
4-Sight (International) Ltd.	617-935-5090	http://www.four-sight.co.uk/4-sight/
Adtran	800-788-5408	http://www.adtran.com/
Advanced Computer Comm.	805-685-4455	http://www.acc.com/
Anderson-Jacobson/CXR	800-433-1425	Not Available

TABLE A.3 Data Networks. (Continued)

Vendor	Phone	Internet Address
Ascend Communications	800-922-0119	http://www.ascend.com/
Ascom Timeplex	800-669-2298	http://www.timeplex.com/
Cisco Systems	800-553-6387	http://www.cisco.com/
Controlware Comm. Systems	908-919-0400	http://www.cware.de/
CoSystems	408-748-2190	Not Available
Digi International	800-551-1259	http://www.dgii.com/
Eicon Technology Corp.	514-631-2592	http://www.eicon.com/
Farallon	510-814-5100	http://www.farallon.com/
Fujitsu ISDN Division	800-228-4736	http://www.fns.com/
Gandalf Systems - Vector Corp.	800-553-5124	http://www.vector.com/gandalf.html
Hewlett-Packard	800-637-7740	http://www.hp.com/
Hitachi America	408-986-9770	http://www.hicam.hitachi.com/
IBM	800-426-2255	http://www.ibm.com/
ISDN Systems	703-883-0933	http://www.infoanalytic.com/isc/index.html
Larscom	408-988-6600	http://www.larscom.com/
Link Technology	215-357-3354	http://www.linkisdn.Inter.net/
Mitel	613-592-2122	http://www.mitel.com/
Motorola ISG		http://www.mot.com/MIMS/ISG/
MPR Teletech	604-293-6047	Not Available
Network Express, Inc.	800-553-4333	http://www.nei.com/
Nortel	800-667-8358	http://www.nt.com
Racal-Datacom	800-333-4143	http://www.racal.com/
Silicon Graphics	800-800-7441	http://www.sgi.com/
Sun Microsystems Computer	800-786-2441	http://www.sun.com/
Symplex	313-995-1555	http://www.symplex.com/

■ VIDEOCONFERENCING

TABLE A.4 Video devices

Vendor	Phone	Internet Address
BT Visual Images	800-778-6288	http://www.bt.com/
Compression Labs Incorporated	800-538-7542	http://www.clix.com/
Intel	800-538-3373	http://www.intel.com/
Nortel	800-992-2303	http://www.nt.com/
PictureTel	800-742-8351	http://www.picturetel.com/

APPENDIX B
Carriers

Tracking down information on the carriers can at times seem frustrating and difficult; it certainly was for me. I hope you find these resources effective for you. They have been gathered from as many sources as I could find and cover a fair portion of the world. As you move out of the United States, you will find a declining number of resources in English, although it was very pleasant and appreciated to find many countries and companies providing extensive English Internet resources.

The resources will begin with the companies in the United States, followed by the long distance companies there, and then by resources from elsewhere in the world. In each area I have provided as much information as is reasonable, but with an emphasis on Internet resources, since this is becoming the fastest way to locate current information on a global basis.

Here every attempt has been made to get you current and effective contact numbers and resource locations. But with the rapid growth of this technology, many companies have been adding or moving their people and programs, making this information more dynamic than we would like. So don't be surprised if you find something is out of date. Try some of the Internet search engines, like Yahoo, Webcrawler, or Lycos to find more current links.

Be wary of the pricing information, it is changing annually or more often in many areas. The contact information should be stabilizing, because many of the vendors recognize the need to support you, and they have established large sales and support groups to answer your questions. They are also putting more effort into creating specialists that they intend to keep working with the service. These people are growing into knowledgeable applications folks that won't reply *"ISDN, what's that?"* or *"3Com, never heard of them or that thing you call an ISDN modem."* But these people are not chartered with helping you design your application, and in many instances, they are prohibited by regulation from giving you specific solution recommendations.

■ THE U.S. LOCAL OPERATING COMPANIES

It has been estimated that while the United States has only 5 percent of the population of the world, its citizens account for 50 percent of the world's data communications resources. In almost every other country in the world you will find one or two companies in the business of telecommunications, but in the United States you have no less than dozens when you include all the regional carriers, thousands with all of the local carriers in rural areas and resellers; with the signing of the telecommunications reform in February of 1996, the number of local exchange carriers will explode over the next few years to thousands more, as many entrepreneurs enter the market on a very small scale.

CARRIERS

How to Reach Them

In TABLE B.1 you will find contact information for all of the major regional carriers. If you do not find your local carrier listed, then contact your telephone business office to see if they can deliver ISDN to you. Figure B.1 is where you will find the territory for each of the regional carriers listed, except for GTE which has its own map in Figure B.2.

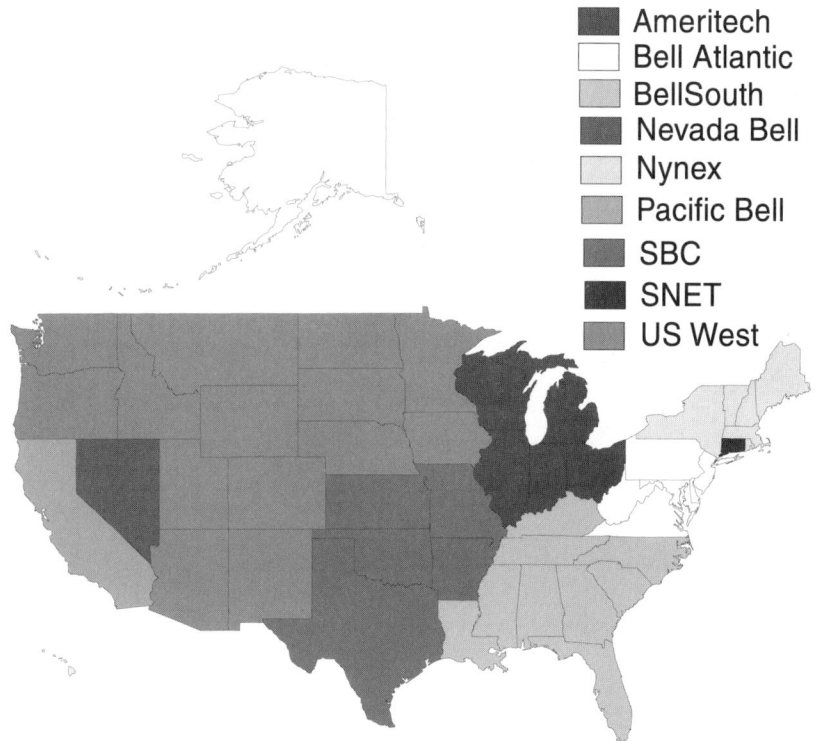

FIGURE B.1 MAJOR REGIONAL OPERATING COMPANIES.

Following the contact information are a series of rate tables covering current offerings, not the complete offerings, but enough of a sample to give you a good feel for the cost of services in a particular part of the country. Please note that many companies have filed tariffs for rate changes, in a number of cases increases, so always check with them for current rates.

A special note for people interested in Alaska, one of my former homes and a treasured memory. The telephone services are delivered through 23 local operating companies and five intrastate long distance carriers. Things are changing there, there is an Alaska 2001 initiative to deal with deregulating the industry and providing the global telecommunications infrastructure they need in the coming years. However, I have chosen not to provide any data for them because of the size and number of companies involved.

TABLE B.1 U.S. operating companies contacts.

Carrier	Contact	Phone	Internet
Ameritech	Ameritech Team Data	800-832-6328	http://www.ameritech.com/products/data/
	Home Prof. & ISDN	800-419-5400	
	Business Prof. & ISDN	800-417-9888	
Bell Atlantic	SDN Sales & Tech.	800-570-4736	http://www.ba.com/isdn.html
	InfoSpeed(tm) Center	800-204-7332	
BellSouth	ISDN Hotline	800-428-4736	http://www.bst.bls.com/products-services/isdn-main.html
Cincinnati Bell	ISDN Service Center	513-566-3282	http://www.cinbelltel.com/
GTE	ISDN Availability	800-483-4926	http://www.gte.com/Cando/cando.html
Nevada Bell	Small Business	702-333-4811	Not Available
	Large Business	702-688-7100	
NYNEX	New York	800-438-4736	http://www.nynex.com/
	New England States	617-743-1333	
Pacific Bell	ISDN Service Center	800-472-4736	http://www.pacbell.com/
	Availability Hotline	800-995-0346	
Rochester Telephone	ISDN Information	716-777-1234	Not Available
SBC	ISDN Information	800-792-4736	http://www.sbc.com/swbell/wct/isdn.html
SNET (Connecticut)	ISDN Sales & Technical Support	800-430-4736	http://www.snet.com/SNET/index.html

TABLE B.1 U.S. operating companies contacts. (Continued)

Carrier	Contact	Phone	Internet
Sprint LTD	ISDN Information	708-768-6043	Not Available
US West	Residence & Home Office	800-898-9675	http://www.uswest.com/isdn/index.html
	Small Business	800-246-5226	
	Facts by Fax	800-728-4929	

Sample Tariffs

The tariff samples here are extracted from rates available in early 1996. With the great mix of services being offered, there is no way to effectively show you all of the tariffs. But this extract should give you a good overview of rates across the country, close enough to help you estimate solution costs.

Where the tariffs define rates based on the capability package selected by the user, I have used the 2B Voice/Circuit-Switched Data + D Packet Data pricing. This rate information is current as of January 1996. In the tables, I have attempted to define usage rates as either measured (M) or flat (F). Measured would be at local toll rates; flat means all usage is at a single price. Both designations indicates you have a choice.

■ Ameritech

The rate structures for this company are fairly complex as they have broken the rates into each capability package. In addition, the rates have a distance extension charge that must be added to them, which I did for this pricing sample. On average the distance extension was $22.50. There are other charges that apply, including options for flat rate service. In each of these states you need to know the capability package you want and the location of your facility before you request a quote.

TABLE B.2 Sample Ameritech ISDN tariffs.

Service/State	Installation	Monthly	Usage
Residential BRI			
Illinois	150.00	50.75	M/F
Indiana	142.00	119.50	V-F/D-M
Michigan	137.00	54.22	M
Ohio	131.50	58.70	M
Wisconsin	128.05	52.15	Local
Business BRI			
Illinois	147.35	61.54	M
Indiana	154.00	119.50	V-F/D-M
Michigan	162.00	60.71	M
Ohio	144.35	69.60	M
Wisconsin	159.65	60.75	Local

Bell Atlantic

The ISDN services in this company are marketed under the name Bell Atlantic IntelliLinQ BRI or PRI. The services must be purchased above the underlying voice services, Centrex or Business Dialtone. There is a program to deliver ISDN anywhere without additional distance charges, when the circuit must be backhauled to a CO capable of providing ISDN. For the PRI tariff, DID services are not included, and the T-1 charges plus the Call-by-Call and Calling Line ID have been added in.

Residential services have entered the trial stage in the following states: Delaware, Maryland, Pennsylvania, Virginia, and West Virginia. Check with the InfoSpeed Center for current information, rates, and availability.

TABLE B.3 Sample Bell Atlantic ISDN Tariffs

Service/State	Installation	Monthly	Usage
Business Line			
Delaware	259.44	56.15	M
Maryland	132.50	44.58	M
New Jersey	196.00	43.36	M
Pennsylvania	287.00	44.73	M
Virginia	109.80	45.83	M
Washington DC	172.50	46.09	M
West Virginia	131.15	58.35	M
Centrex BRI			
Delaware	157.00	45.20	F/M
Maryland	130.50	44.25	F/M
New Jersey	233.25	38.67	F/M
Pennsylvania	123.00	44.68	F/M
Virginia	100.80	47.25	F/M
Washington DC	155.25	43.25	F/M
West Virginia	109.90	56.00	F/M
Business PRI			
Delaware	1400.00	750.00	M
Maryland	1400.00	650.00	M
New Jersey	1400.00	600.00	M
Pennsylvania	1400.00	650.00	M
Virginia	1400.00	600.00	M
Washington DC	1400.00	600.00	M
West Virginia	1400.00	950.00	M

■ BellSouth

Services in this territory are marketed as ESSX ISDN service, ISDN Individual Business Service, and ISDN Individual

Residence Service. ESSX ISDN is the Centrex offering and Mega-Link ISDN is the PRI offering. The PRI does not include the charges for the underlying voice services including: Hunt, Rotary, DID, etc.

TABLE B.4 Sample BellSouth ISDN Tariffs

Service/State	Installation	Monthly	Usage
Residential BRI			
Alabama	165.00	66.10	F
Kentucky	165.00	56.55	F
Louisiana	165.00	65.00	F
Mississippi	165.00	62.01	F
Tennessee	0.00	26.00	F
Florida	165.00	53.65	F
Georgia	165.00	59.90	F
North Carolina	165.00	71.51	F
South Carolina	165.00	57.90	F
Business BRI			
Alabama	165.00	101.00	F
Kentucky	165.00	101.00	F
Louisiana	165.00	101.00	F
Mississippi	165.00	101.00	F
Tennessee	165.00	101.00	F
Florida	165.00	101.00	F
Georgia	165.00	101.00	F
North Carolina	165.00	99.50	F
South Carolina	165.00	101.00	F
Centrex			
Alabama	165.00	101.00	F
Kentucky	165.00	101.00	F
Louisiana	165.00	101.00	F
Mississippi	221.00	46.05	F
Tennessee	221.00	19.00	F

TABLE B.4 Sample BellSouth ISDN Tariffs (Continued)

Service/State	Installation	Monthly	Usage
Florida	221.00	35.80	F
Georgia	168.00	26.10	F
North Carolina	163.00	35.10	F
South Carolina	221.00	41.80	F
Business PRI			
Alabama	1177.00	1335.00	M
Kentucky	956.00	1462.00	M
Louisiana	1181.00	1285.00	M
Mississippi	1156.00	1315.00	M
Tennessee	1201.00	1305.00	M
Florida	1156.00	1400.00	M
Georgia	1163.00	1317.00	M
North Carolina	1163.00	1317.00	M
South Carolina	1126.00	2034.00	M

■ Cincinnati Bell

There are three services offered by Cincinnati Bell, ISDN Business — Flex Line, ISDN Centrex — 2000/90+, and ISDN PRIME Advantage. In Ohio there are two rate structures for BRI, both flat and measured.

TABLE B.5 Sample Cincinnati Bell ISDN tariffs.

Service/State	Installation	Monthly	Usage
Centrex BRI			
Ohio – Flat	32.67	112.93	M/F
Ohio - Measured	Request	Request	M/F
Kentucky	32.67	112.93	M/F
Business BRI			
Ohio – Flat	72.59	112.05	M/F
Ohio – Measured	72.59	70.05	M/F
Kentucky	72.59	70.05	M/F

■ GTE

This company provides service in the states shown in Figure B.2; 20 states today. ISDN has been offered in 15 of those states as of late 1995, with the last five coming on-line in 1996. GTE has been behind many of the other regional operating companies in offering ISDN services, but today they have achieved just over 50 percent availability, and they are aggressively pursuing digital technologies. In a trial that began in early 1996, GTE became the first carrier to test ADSL to deliver services. While making that commitment to ADSL, they also moved to establish flat rate tariffs to stimulate the ISDN market and committed to AT&T to purchase and install a significant number of new ISDN circuits.

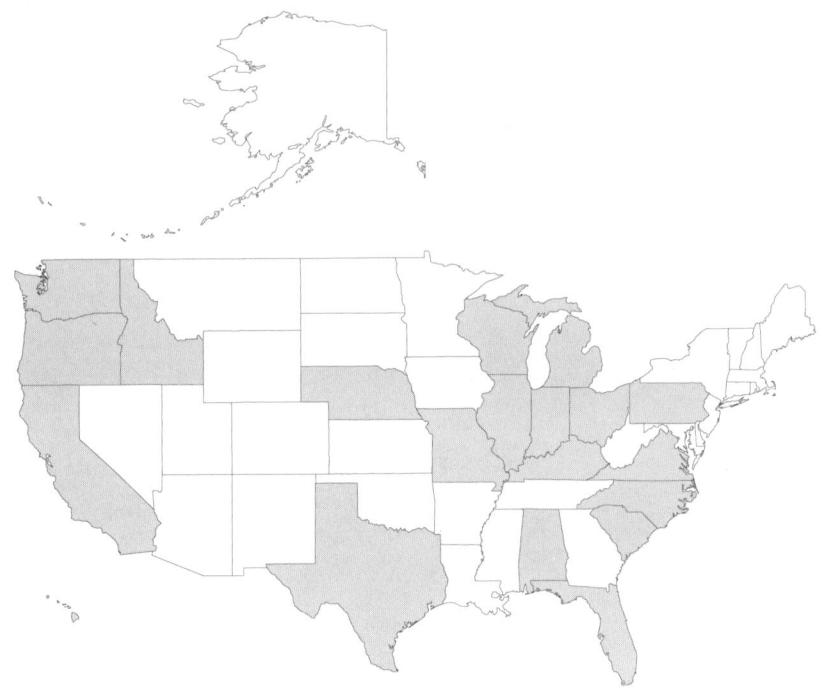

FIGURE B.2   GTE OPERATING AREAS, INCLUDES HAWAII.

The Centrex service is offered under the Digital CentraNet brand name and has a minimum number of lines requirement. The

Washington flat rate tariff has limited usage categories, cutting the costs up to 50 percent for 25 hours of use. The PRI tariffs are complex, most carry a separate installation charge for B channel options, like DID, DOD, Call-by-Call, and have a per-channel monthly recurring cost. All tariffs listed do not include these other services, which can add $200 per service in installation fees and $4 to $51 per channel in monthly recurring cost.

TABLE B.6 Sample GTE ISDN tariffs.

Service/State	Installation	Monthly	Usage
Residential BRI			
California	134.61	37.50	M
Florida	135.00	32.50	M
Hawaii	121.00	38.90	M
Illinois	83.00	46.96	M
Indiana	116.30	52.88	M
Kentucky	108.45	53.39	M
North Carolina	100.85	67.30	M
Ohio	99.35	51.23	M
Oregon	129.00	35.57	M
Pennsylvania	100.00	44.95	M
Texas	171.00	42.10	M
Virginia	74.25	40.75	M
Washington	123.00	76.70	F
Business BRI			
California	205.72	49.22	M
Florida	148.90	48.67	M
Hawaii	121.00	62.60	M
Illinois	106.00	56.98	M
Indiana	124.60	81.68	M
Kentucky	109.85	85.89	M
North Carolina	111.50	69.80	M
Ohio	106.35	62.89	M

TABLE B.6 Sample GTE ISDN tariffs. (Continued)

Service/State	Installation	Monthly	Usage
Oregon	139.00	48.00	M
Pennsylvania	111.00	53.77	M
Texas	101.50	44.60	M
Virginia	74.25	63.14	M
Washington	147.00	95.80	F
CentraNet			
California	235.72	58.22	M
Florida	173.90	77.81	M
Hawaii	146.00	94.00	M
Illinois	131.00	71.50	M
Indiana	149.60	99.50	M
Kentucky	134.85	98.25	M
Michigan	30.00	43.05	M
North Carolina	125.85	112.25	M
Ohio	131.35	93.50	M
Oregon	169.00	99.75	M
Pennsylvania	136.00	84.00	M
Texas	135.50	52.10	M
Virginia	99.25	90.49	M
Washington	177.00	85.25	M
PRI			
California	1358.80	782.00	M
Florida	961.90	612.00	M
Hawaii	526.00	817.00	M
Illinois	1349.00	703.40	M
Indiana	1190.66	687.00	M
Kentucky	440.25	1096.40	M
North Carolina	539.55	1320.80	M
Ohio	1148.25	689.00	M
Oregon	1174.00	563.21	M

TABLE B.6 Sample GTE ISDN tariffs. (Continued)

Service/State	Installation	Monthly	Usage
Pennsylvania	1430.00	631.00	M
Texas	405.56	562.00	M
Virginia	768.25	637.00	M
Washington	583.00	492.00	M

■ Nevada Bell

The ISDN centrex services are priced by volume with discounted rates available to sites with as few as seven lines. The example is for one line.

TABLE B.7 Sample Nevada Bell ISDN tariffs.

Service/State	Installation	Monthly	Usage
Business BRI			
Nevada	185.00	80.43	M

■ NYNEX

The tariffs require that you add all channel services to existing telephone facilities. When the CO serving your site is not capable of ISDN, ISDN can be installed under their *Virtual ISDN* service, for a flat fee of $10 per month for each BRI circuit. If the distance to your site is over 18,000 ft. you may be able to get ISDN, but this service is currently under investigation. Outside of New York, 2B+D services are not available, so the tariff sample is a 1B+D voice/circuit-switched data rate. Residential rates are the same as the business rates.

For both the BRI and PRI level of services, there is an End User Common Line Charge per BRI line or per B channel on the PRI circuit, and it averages $5.80. The PRI tariffs are for installation within 1/2 mile of the CO, distance extension charges apply for each 1/2 mile after that.

TABLE B.8 Sample Nynex ISDN tariffs.

Service/State	Installation	Monthly	Usage
Business BRI			
New York	195.00	16.00	M
Massachusetts	143.02	32.60	M
Maine	151.00	49.20	M
Vermont	126.00	59.50	M
New Hampshire	135.00	41.35	M
Rhode Island	139.61	43.31	M
Centrex			
New York	195.00	46.00	M
Massachusetts	90.00	56.00	M
Maine	150.00	48.75	M
Vermont	245.00	59.07	M
New Hampshire	135.00	46.00	M
Rhode Island	150.00	49.00	M
Business PRI			
New York	1125.00	927.25	M
Massachusetts	1697.38	715.00	M
Maine	1625.00	1010.00	M
New Hampshire	2395.00	960.00	M

■ Pacific Bell

FasTrak is the brand for ISDN services in the California area, with Centrex, Business, Home, and Primary Rate Services. Service is available from any location through the backhaul arrangement and loop length concerns are not an issue because Pacific Bell will install digital repeaters at no charge.

For PRI there is a T-1 access charge in addition to the ISDN rates, included in this estimate. Optional services, like DID, DOD,

WATS, etcetera, are subscribed to in addition to the PRI rates. Lines can be configured for Call-by-Call or fixed capacity, or maximum capacities within the Call-by-Call model.

Usage charges are based on the call distance from your site. The home ISDN product carries a measured rate on all calls Monday–Friday, 8 AM to 5 PM, then it returns to the normal flat rate for local calling.

Pacific Bell has filed for rate increases for all ISDN products. This is based on the need to recover more costs from the product and to provide enough revenue to resell the ISDN products to other carriers. With the deregulation of the industry, the California PUC instructed Pacific Bell to make ISDN a product that can be resold, so other companies can purchase the services in bulk and offer them to you under their own label. Average rate increase for the BRI services will be $8 per month; usage remains the same.

TABLE B.9 Sample Pacific Bell ISDN tariffs.

Service/State	Installation	Monthly	Usage
Residential BRI			
California	34.75	24.50	F/M
Business BRI			
California	70.75	25.93	M
Centrex			
California	70.00	31.65	M
Business PRI			
California	750.00	570.00	M

■ SBC

SBC has made available many of the features of ISDN. Installation services allow you to put in ISDN on extended local loops (greater than 18,000 ft.) and from an ISDN capable CO within your exchange area without additional charges. The services are known as DigiLine, SmartTrunk, and Plexar w/ISDN. Services are configured according to the capability packages. Not all states have ISDN tariffs; some services are still undergoing approval, while Arkansas & Oklahoma are just getting started with this process. Centrex is currently a custom contract negotiated for each installation.

The PRI services include the facility charge, the T-1, and flat rate usage. For this sample I included the optional Call-by-Call service, but not the Calling Line Identification.

TABLE B.10 Sample SBC ISDN tariffs.

Service/State	Installation	Monthly	Usage
Business BRI			
Kansas	405.50	50.50	M
Missouri	400.00	50.50	M
Texas	485.00	48.00	F/M
Business PRI			
Arkansas	2100.00	1898.00	F
Kansas	1800.00	1765.00	F
Missouri	1850.00	1938.00	F
Oklahoma	2000.00	1667.00	F
Texas	1275.00	1049.00	F

■ SNET

The service for BRI is marketed as Digital Enhancer and provides extensive voice and data services. The tariff is built around the specific Electronic Key Telephone Services functions of ISDN, with specific rates for the number of features per button count, etcetera. The rate listed is for the basic service.

TABLE B.11 Sample SNET ISDN tariffs.

Service/State	Installation	Monthly	Usage
Residential BRI			
Connecticut	265.00	50.00	M

■ US West

This carrier emphasizes that you should check their Web site for information on the current tariffs. From their web site the following information was gathered.

TABLE B.12 Sample US West ISDN tariffs.

Service/State	Installation	Monthly	Usage
Basic BRI			
Washington	85.00	63.00	F
South Dakota	80.00	84.00	
Oregon	110.00	69.00	F/M
Arizona	110.00	69.00	F/M
Colorado	67.00	60.00	
Minnesota	110.00	69.00	F/M
Business PRI			
Washington	3095.00	2234.00	F
Oregon	4002.00	2156.00	F
Minnesota	5598.00	1982.00	F
Arizona	3512.00	2235.00	F
Colorado	3512.00	2198.00	F

■ THE LONG DISTANCE COMPANIES

Your long distance rates will apply to all calls made outside the local calling area from your local ISDN circuit or from your dedicated ISDN circuit from the long distance carrier. TABLE B.14 is just for AT&T to give you a feel for both U.S. and international rates. Other carriers will be comparable, and all of these rates will be subject to the pricing agreements you have with the carrier.

TABLE B.13 Long distance carrier contacts.

Carrier	Contact	Phone	Electronic
AT&T	AT&T Front End Center	800-222-7956	http://www.att.com/
LDDS	ISDN Product Management	201-804-6970	http://www.wcom.com/
MCI	ISDN Availability	800-624-4736	http://www.mci.com/
Sprint	Nancy Johnson	913-624-4308	http://www.sprint.com/

The AT&T domestic rates are based on mileage, total bandwidth, and time of day. All calls are billed in a 30 second minimum with 6 second increments.

TABLE B.14 Sample AT&T ISDN tariffs.

Service/State	30 secs	6 seconds	Hour
Domestic ISDN Day Rate			
56/64 Kbps 0-55 Miles	0.1395	0.0119	14.44
56/64 Kbps 1911-3000 Miles	0.2000	0.0240	28.96
384 Kbps 0-55 Miles	0.4813	0.0403	24.46
384 Kbps 1911-3000 Miles	0.6843	0.0809	48.82
Selected Countries			
Japan 56/64 Kbps Peak	1.2450	0.2490	298.80
Japan 56/64 Kbps Off-Peak	0.9450	0.1890	226.80
Japan 384 Kbps Peak	4.980	0.9960	597.60
Japan 384 Kbps Off-Peak	3.780	0.7560	453.60
Australia 56/64 Kbps Peak	1.2450	0.2490	298.80
Australia 56/64 Kbps Off-Peak	0.9450	0.1890	226.80

■ GOING INTERNATIONAL

While the high speed telecommunication's revolution is getting off the ground in the United States, other countries have been embracing ISDN technology for years, with incredible levels of implementation in some of those countries. While the physical size of a country can greatly enhance their ability to deliver an innovative tariff, it shouldn't diminish the accomplishment of wide-scale ISDN deployment.

For example, Deutsche Telekom has an on-line service, T-Online, which has country wide access to its services by ISDN. The company has installed over 2.8 million basic ISDN channels. The boom for ISDN is just really getting of the ground in the United States, so don't be surprised when there's no wonder in the eyes of someone that is used to wide spread access to this technology, like in Germany.

I begin the resources internationally with TABLE B.15, where numerous countries are listed with either voice, fax, or electronic contacts for each telecommunications company.

TABLE B.15 Non-U.S. local carrier contacts.

Country	Carrier	Phone	Electronic
Australia	Telstra		http://www.telstra.com.au/
Austria	PTT	43 1 31 313 663F	Not Available
Belgium	Belgacom	32 2 202 47 23F	http://www.belgacom.be/
Bermuda	Bermuda Telephone	1 809 292 1192F	Not Available
Brazil	Embratel		http://www.embratel.net.br/
Canada	Stentor	800 578 4736F	http://www.stentor.ca
Cyprus	Cyprus Telecom	357 249 4155F	Not Available
Denmark	Telekom Denmark	45 80 209 2076F	http://
Dominican Republic	Codetel		http://www.codetel.net.do/codetel/mencod.html
Finland	The Association of Telephone companies in Finland	358 0 228 111V 358 0 605 531F	http://www.tpo.fi/english/finland/finltcom.html
Finland	Telecom Finland		Not Available
France	France Telecom	33 1 43 42 97 77F	http://www.francetelecom.com/
Germany	Deutsche Telekom	49 651 130 2305F	http://www.dtag.de/dtag/telekom_.html
Greece	Hellenic	30 1 805 2064F	Not Available
Hong Kong	Hong Kong Telecom	852 2824 0077F	http://www.hkt.net/
Iceland	Telecom Iceland	354 550 6209F	Not Available

TABLE B.15 *Non-U.S. local carrier contacts. (Continued)*

Country	Carrier	Phone	Electronic
Ireland	I. E. Telecom Eireann	353 1 677 4941F	http://www.broadcom.ie/telecom/dupjmc/index.html (gets you close, they're not on the Internet)
Israel	BEZEQ The Israel Telecommunication Corp. Ltd.	972 3510 0696F	http://www.bezeq.co.il/eindex.html
Italy	Telecom Italia	396 3687 4531F	http://www.telecomitalia.interbusiness.it/Telecom/en/index.html
Japan	IDC	813 5820 5371F	Not Available
Japan	ITJ	813 5565 5330F	Not Available
Japan	KDD	813 5225 7473F	http://www.kdd.co.jp/
Japan	NTT	44 171 496 0390F	http://www.ntt.jp/
Luxembourg	P & T Luxembourg	352 2729 0700F	Not Available
Malaysia	Telekom Malaysia	603 254 3435F	Not Available
Netherlands	PTT Telecom	31 70 34 39 74F7	Not Available
New Zealand	Telecom New Zealand	64 4 498 9117F	Not Available
Norway	Telenor Int.	47 22 11 03 35F	http://www.telenor.no/
Philippines	Telephone Company	63 812 4936F	Not Available
Portugal	Telecom Portugal	351 1 572474F	Not Available
Russia	Comstar Telecom	7 095 956 2205F	Not Available
Russia			http://mp.aha.ru/koi/isdn/
Singapore	Singapore Telecom	65 733 3008F	http://www.singtel.com/index.html
South Africa	Telecom South Africa Ltd.	27 12 311 1146V	http://www.telkom.co.za/
Spain	Telefonica de Espana	34 1584 9558F	Not Available
Sweden	Telia AB	46 8 713 73 62F	http://www.telia.se/index_en.html
Switzerland	Swiss Telecom PTT	41 31 338 8526F	http://www.vptt.ch/
Taiwan	ITA	886 2344 2693F	Not Available
United Arab Emirate	Etisalat	9712 324499F	Not Available
United Kingdom	British Telecom	44 117 927 4218F	http://www.bt.net/

TABLE B.15 Non-U.S. local carrier contacts. (Continued)

Country	Carrier	Phone	Electronic
United Kingdom	Mercury	44 1344 726300F	Not Available

Tariff Examples

Rate information included in the next example is shown in local currency, rather than converted US dollars, exchange rates are still more volatile than the ISDN rates. These figures are good as of February 1996 and should be used only as an estimation of your costs. At a number of sites I found indications that rates were changing, so remember to confirm your plans by contacting the carrier directly.

■ Telstra

This is an interesting tariff for Internet access in Australia. As expected, there are the traditional installation and monthly usage charges, but monthly usage is governed by unusual characteristics. Usage is defined as the amount of inbound traffic delivered to a line for the entire month. The total bandwidth used on this inbound transmission is then divided by the available bandwidth on the pipe, and then this average usage rate is used to look up the monthly charge

This model is used to promote the delivery of content onto the Internet, not the consumption. It is an interesting model to spur electronic commerce. Take a look at the following example to understand how the charges would apply. If you need large download capacities, this tariff can get expensive, with the top charge for a 64 Kbps circuit being $9,000 per month.

The rates listed in TABLE B.16 are only a brief excerpt of the seven access levels, ranging from 64 Kbps to 10 Mbps, that are available. Also, the usage sampling in the actual rates go up in

> You have an Internet connection on a single BRI and have only dedicated one channel to this traffic. Your objective is to receive files from other locations on a daily basis, approximately 30 MB each day.
> Here's the formula to compute your usage percentage:
> Bits Received(BR) / Total Capacity of Line for the period(TC)
> For our example, covering 30 days:
> 7,549,747,200 BR / 165,888,000,000 TC = 4.5%
> From the tables for this tariff this would be $2,000 for usage, which is also the minimum usage charge.

5 percent increments, and I have only included three to show the range of the tariff.

TABLE B.16 Sample Telstra ISDN rates.

Charge	64 Kbps	128 Kbps	2048 Kbps
Installation Options			
Router Port	$3,500	$3,500	$3,500
ISDN Router, Single Port	$5,100	$8,750	N/A
ISDN Router, Multiport	N/A	$10,200	N/A
Monthly Usage			
0 to 5%	$2,000	$3,100	$12,500
>50 to 55%	$3,500	$5,500	$23,500
>95 to 100%	$9,000	$14,167	$95,167

■ France Telecom

In looking through the tariffs for France Telecom I found both domestic and International services for ISDN. I have included examples of both tariffs so you can get a feel for the direct cost of connection from outside the country, and what your charges would be if you used the Internet as a go-between for your datacommunication needs. The name of the service in France is Numeris.

The international tariff is interesting, because it includes usage rates based on zones, much like the United States practice of zoned rates in the local calling area, only this covers the world. The basic tariff is shown in TABLE B.17, with rates that were valid as of June, 1994, although this is still the post on their web site. To these prices you need to add the French VAT of 20.6 percent, a hefty increase in costs.

TABLE B.17 Sample France Telecom international ISDN rates.

Charge	2B + D	30B + D
Installation	675 FF	4,200 FF
Monthly Charges	200 FF	104 FF / Outgoing Channel 32 FF / Incoming Channel Minimum of 960 FF
Country Zones - Rate FF/Min.		
Zone A Belgium, Germany, Italy, Luxembourg, the Netherlands, Spain, Switzerland, United Kingdom	4.00 FF	4.00 FF / Channel
Zone B Denmark, Ireland, Portugal	4.82 FF	4.82 FF / Channel
Zone C Finland, Norway, Sweden	6.77 FF	6.77 FF / Channel
Zone E Canada, United States	7.89 FF	7.89 FF / Channel
Zone I Australia, Hong Kong, Japan, New Zealand, Singapore	14.76 FF	14.76 FF / Channel
Zone L South Africa	18.45 FF	18.45 FF / Channel

One of the interesting notes on the service in France is that the D-channel packet service connects you to the France Telecom-Transpac network, opening the possibility of having 64 Kbps packet service to anywhere in France.

The domestic rates for installation and monthly recurring charges are the same as they are for international, with one added service, Grouped Basic Rate Interface First Access, which carries a 300 FF monthly usage charge. Phone call tolls are then added to these charges plus the 20.6 percent French VAT. What's different about this tariff is that the calls are distinguished between digital to analog and digital to digital. In TABLE B.18 I have included the cost for only the local and most distant calls, there are distance brackets in-between these distances.

TABLE B.18 Sample France Telecom domestic ISDN rates.

Service Area	Peak	Reduced 1	Reduced 2	Reduced 3
Digital to Analog				
Local Calls	.21 FF/min	.14 FF/min	.10 FF/min	.07 FF/min
> 100 Km	1.75 FF/min	1.23 FF/min	.87 FF/min	.615 FF/min
Digital to Digital				
Local Call	.51 FF/min	.36 FF/min	.26 FF/min	.18 FF/min
>100 Km	2.46 FF/min	1.75 FF/min	1.23 FF/min	.88 FF/min

APPENDIX C
Other Resources

It is always frustrating to have a bunch of ideas, but then be unable to find more information or find the people to actually help you purchase and implement them. This chapter of the book is dedicated to giving you direction to the resources that will be of benefit. This is by no means an exhaustive list — it's almost impossible to do that. But it is a good starting point to build from.

I start with on-line resources from the Internet and other services, like CompuServe, because these sources provide you with fairly current information. More importantly, they give you access to people all over the globe who are working with the technology, people like you trying to implement it, and the vendors who are trying to sell it to you. On to the books, because they are still one of the best forms of reference material for concepts and is the reason I decided to write this book and have it added to other people's lists. The books I have included cover

ISDN and other topics in greater depth if you are inclined to dig deeper into the technology.

■ ON-LINE RESOURCES AND USERS GROUPS

Most of the resources listed here are on the Internet; and with the nearly universal access to this resource, this is the logical place for you to start. The first resource listed is Dan Kegel's web site, where perhaps the most extensive list of on-line resources is kept. This web site is one of the places I frequent because it can be counted on to be relatively current, and certainly exhaustive. But as with all Internet resources, more abound that can be found only by crawling around the Internet, and those have been listed also.

Users Groups

Always an excellent source of information, users groups provide two types of resources for you. TABLE C.1. lists a number of the users groups that are accessible from the Internet. The most important thing is the contact with others doing the same type of work that you are, and solving problems in ways you haven't thought about. There is nothing more powerful than a room full of users to find a solution or help a person out. Maybe that is why so many companies have provided the many on-line forums to let users help each other, reducing everyone's support costs, but more importantly, getting more practical information in the process.

TABLE C.1. Users groups.

Organization	Notes	Web Site/E-mail	Phone
California ISDN User Group	Has a number of information pages.	http://www.ciug.org/ info@ciug.org	415-241-9943 V 415-753-6617 F
Florida ISDN User Group	New homepage, under construction as the book went to press.	http://www.fiug.org mengel@packet.net	813-588-1206 V 813-588-1277 F
New York ISDN User Group	Basic information page with meeting information.	http://www.users.interport.net/ ~digital/about.html	212-944-5400 V 212-944-5410 F

OTHER RESOURCES

TABLE C.1. Users groups. (Continued)

Organization	Notes	Web Site/E-mail	Phone
Texas ISDN User Group	Good home page, lots of information. Also mirrors Dan Kegel's web pages.	http://www.crimson.com/isdn/ info@tiug.org	
Indonesia ISDN & IN User Forum	Extensive site, well organized.	http://www.idola.net.id/i3uf/ jasnita@idiola.net.id	62 21 7201221 V 62 21 7201226 F
Pacific Region ISDN/Data User Forum	This is the group within the Pacific Telecommunications Council to develop applications and services within the Pacific Region.	http://www.ptc.org:80/PRIDUF/ gholt@aloha.com	510-736-6924 V 510-855-1428 F
North American ISDN User's Forum	The most comprehensive site I found dedicated to ISDN.	http://www.niuf.nist.gov/misc/niuf.html niuf@nist.gov	301-975-2937 V 301-926-9675 F
South African ISDN Forum	Is a relatively new home page, still under development.	http://www.saif.org.za/ siaf@siaf.org.za	

General Resources

The next table lists a number of on-line sites I have found valuable. I haven't tried to list all the sites I found. Instead I wanted to point you in the direction of the ones with comprehensive or unusual coverage, which will draw you even further into the Internet and on-line services. I have highlighted one site, Dan Kegel's, because of the incredible effort and strength of information found there on ISDN and its competing technologies. If you go nowhere else to find anything, go to this site.

TABLE C.2. On-line resource sites.

Site	Address	Notes
Alta Vista	http://www.altavista.digital.com/	Search service.
Bellcore	http://www.bellcore.com/ISDN/ISDN.html	ISDN information.

TABLE C.2. On-line resource sites. (Continued)

Site	Address	Notes
Chico Junior High School	http://www.chicojr.chico.k12.ca.us/	Application example.
CICAT ISDN Products and Service	http://www.cicat.com/cicat/	Vendors' products.
Cook, Jeff	http://www.infinet.com/~isdnjeff/	ISDN information
Dan Kegel's Web Site	**http://alumni.caltech.edu/~dank/isdn/index.html**	**Probably the best site on the Internet.**
Dateline ISDN	http://www.insgroup.com/nis/register.htm	On-line news service.
David Ginsberg's ISDN Page	http://www.caprica.com/~dginsberg/isdn.html	Macintosh ISDN information.
DejaNews	http://www.dejanews.com/	News service.
Digital Velocity ISDN	http://www.isdn.nortel.net/	Nortel's information site.
FAQ	http://www.cis.ohio-state.edu/hypertext/faq/usenet/isdn-faq/faq.html	Frequently Asked Questions File — aging but still excellent.
FAQ	http://www.multithread.co.uk/isdnfaq.htm	United Kingdom ISDN information.
Goodman's Book Marks: International Telecom	http://www.wp.com/goodmans/intl.html	International telecommunications information.
ISDN and Home Networks	http://www.scruznet.com/~springer/wires/	ISDN information.
ISDN InfoCentre	http://www.isdn.ocn.com/	ISDN information.
ISDN Quick; Availability/Pricing Database	http://www.insgroup.com/nis/qwikquot.htm	National pricing and availability information.
ISDNsights	http://techweb.cmp.com/nwc/isdn/	ISDN information.
Inktomi	http://inktomi.cs.berkeley.edu:1234/	Search service.
Interdisciplinary Telecommunications Program	http://morse.colorado.edu/	Training organization.
Jeff Frohwein	http://fly.HiWAAY.net/~jfrohwei/isdn/	ISDN Information

OTHER RESOURCES

TABLE C.2. On-line resource sites. (Continued)

Site	Address	Notes
Lycos	http://lycos.cs.cmu.edu/	Search service.
Open Text	http://search.opentext.com/	Search service.
Shea Communications	http://www.SheaComm.com/	ISDN and publishing.
Usenet	news:comp.dcom.isdn	Internet newsgroup.
WebCrawler	http://www.webcrawler.com/	Search service.
Woodbridge ISDN Housing Development	http://www.mother.com/woodbridge/	Application example.

Resellers, Software, Consultants

These companies all provide expertise in designing, installing, and operating ISDN solutions. While they may not be local, if they can't solve your problems, they should be able to point you in the right direction.

TABLE C.3. Hardware, software, consultants.

Site	Address	Notes
AHK & Associates	http://www.value.net/ahk/html/	Consultants and products.
Avatar Systems Ltd.	http://www.avsysltd.com/	Consultants.
BF Datacom	http://www.bfdatacom.com/	Consultants.
Consensus Informatik AG	http://www.consensus.ch/	Consultants — Germany.
Data Accessories of Texas	http://www.phoenix.net/~dtaacces/	Equipment and network design.
DIGINET, Inc.	http://www.diginetinc.com/	Consultants and products.
Dittberner Associates	http://www.dittberner.com/	Consultants.
Dillon Technology Group, Inc.	http://www.fentonnet.com/dillon/	Consultants and products.
ETI Group	http://www.neosoft.com/~eti/	Wireless ISDN.
Gitter, Orlowsky & Associates	http://www.shadow.net/~gitter/	Consultants and products.
Helfrich Company	http://WWW.HELFRICH.COM/	Consulting and test equipment.
ISDN for Linux	http://www.ix.de/ix/linux/linux-isdn.html	Software.

TABLE C.3. Hardware, software, consultants. (Continued)

Site	Address	Notes
Link Technology, Inc.	http://linkisdn.inter.net/linkisdn/	Hardware and software developers for ISDN.
netCS	http://www.netcs.com/	Unix Solutions — Germany.
Odin TeleSystems Inc.	http://www.OdinTS.com/	Test equipment.
Powercom and One Com	http://www.powercom.com/	Consultants.
Results From Technology!	http://ourworld.compuserve.com/homepages/roblee/	Consultant.
Synapse, Inc.	http://www.isdnshop.com/	Consultants and product.
Telecom Consultants Inc.	http://www.tcinetwork.com/	Consultants.
Telcom Systems Services, Inc. (TSS)	http://telsyserv.com/	Telephone systems reseller.
Valueware	http://www.kwik-link.com/kwik-link/c/Vcn41.htm	Hardware and software.

Broadcast Audio

This was a brief set of resources I found on the broadcast industry using ISDN. The first and last entries are loaded with contact information.

TABLE C.4. Broadcast ISDN resources.

Site	Address	Notes
Broadcast ISDN User Guide and Directory	http://www.sms.co.uk/isdn/	Information from the book.
Future Sound of London [gatech.edu]	http://ulc199.residence.gatech.edu/guyjr/fsol.html	Application example.
Intraplex, Inc.	http://www.intraplex.com/	Equipment.
Telos Systems	http://www.zephyr.com/	Equipment.
Zephyrlist	www.tiac.net/users/jcrose/audiobahn.html	Services and users.

On-line Services Providers

You will have to pay extra to access these services, but it is really worth the money. Each service has a forum dedicated to ISDN or any of the application areas that ISDN supports. Most of the services have a free trial period to let you find the areas of interest and judge the quality of the content.

TABLE C.5. On-line services providers.

Company	Internet Address
America Online	http://www.aol.com/
Compuserve	http://www.compuserve.com/
Microsoft Network	http://www.msn.com/
Prodigy	http://www.prodigy.com/

Books

ISDN Technology

Bryce, James Y., *Using ISDN,* special edition (Indianapolis, IN: Que, 1995).

Tittel, Ed, and Steve James, *ISDN Networking Essentials* (Chestnut Hill, MA.: AP Professional, 1996).

Summers, Charles K., *ISDN Implementor's Guide* (New York, NY.: McGraw-Hill, Inc., 1995).

Hopkins, Gerald L., *The ISDN Literacy Book* (Reading, MA.: Addison-Wesley Publishing Company, 1995).

Angell, David, *ISDN For Dummies* (Foster City, CA.: IDG Books Worldwide, Inc. 1995).

Moore, Martin L., *ISDN Strategies* (Foster City, CA.: IDG Books Worldwide, Inc., 1995).

Kessler, Gary C., *ISDN* (New York, NY.: McGraw-Hill, Inc., 1990).

Hardwick, Steve, *ISDN Design* (San Diego, CA.: Academic Press, Inc., 1989).

Stallings, William, *Advances in ISDN and Broadband ISDN* (Los Alamitos, CA.: IEEE Computer Society Press, 1992).

Smouts, M., *Packet Switching Evolution from Narrowband to Broadband ISDN* (Boston, MA.: Artech House, 1991).

Network & Systems Design

Spohn, Darren L., *Data Network Design* (New York, NY.: McGraw-Hill, Inc., 1993).

Coad, Peter and Edward Yourdon, *Object-Oriented Analysis* (Englewood Cliffs, NJ.: Prentice-Hall, Inc., 1991,1990).

Morris, Daniel and Joel Brandon, *Re-engineering Your Business* (New York, NY.: McGraw-Hill, Inc., 1993).

Jain, Bijendra N. and Ashok K. Agrawala, *Open Systems Interconnection* (New York, NY.: McGraw-Hill, Inc., 1993).

Freedman, Daniel P. and Gerald M. Weinberg, *Handbook of Walkthroughs, Inspections, and Technical Reviews* (New York, NY.: Dorset House Publishing, 1990).

Gause, Donald C. and Gerald M. Weinberg, *Are Your Lights On?* (New York, NY.: Dorset House Publishing, 1990).

Spragins, John D. with Joseph L. Hammond and Krzysztof Pawlikowski, *Telecommunications Protocols and Design* (Reading, MA.: Addison-Wesley Publishing Company, 1991).

McClimans, Fred J., *Communications Wiring and Interconnection* (New York, NY.: McGraw-Hill, Inc., 1992).

People and Project Skills

Graham Scott, Gini, *The Empowered Mind* (Englewood Cliffs, NJ.: Prentice Hall, 1994).

Kushner, Malcom, *The Light Touch* (New York, NY.: Simon and Schuster, 1990).

Wonder, Jacquelyn and Priscilla Donovan, *Whole-Brain Thinking* (New York, NY.: William Morrow and Company, Inc. 1984).

Dawson, Roger, *The Confident Decision Maker* (New York, NT.: William Morrow and Company, Inc., 1993).

Sheth, Jagdish N. and S. Ram, *Bringing Innovation to Market* (New York, NY.: John Wiley & Sons, 1987).

OTHER RESOURCES

Quick, Thomas L., *Unconventional Wisdom* (San Francisco, CA.: Jossey-Bass Inc., 1989).

Weeks, Dudley, *The Eight Essential Steps to Conflict Resolution* (New York, NY.: G.P. Putnam's Sons, 1992).

Shelton, Ken, ed., *In Search of Quality* (Provo, UT.: Executive Excellence Publishing, 1995).

Appendix D: The CD-ROM

Included with this book is a CD-ROM loaded with some great software and resources for you to use. This book has been created in the Adobe Acrobat format so you can search through the book using the text indexing features of the Acrobat software. This will help you return quickly to areas of the book when you are working on your designs. The other major document included is the North American ISDN Users Forum application catalog. This book was put together by many industry experts and was converted to the Acrobat format to make it readily available to everyone. After these resources, I have included a demonstration version of the netViz product for diagramming, additional forms or templates to supplement what is in the book, and of course, the Acrobat software to read the books.

■ CD-ROM Organization

The CD-ROM has been mastered to be read by Windows, Macintosh, UNIX, and DOS systems. Most of the software runs on the Windows platform. The Adobe Acrobat software will operate in all four environments, so both the book and the applications catalog are available to you. The netViz demo software only runs in Windows.

The directory structure for the CD-ROM is shown in TABLE D.1. All directories are contained in the ISDN-CD subdirectory.

TABLE D.1 CD-ROM directories.

Level 2	Level 3	Purpose
ACRO_V1	ACRODOS	Contains the Acrobat software for DOS 3.3 or greater systems.
	ACROSGI	Contains the Acrobat software for SGI IRIX systems.
ACRO_V2	ACROMAC	Contains the Acrobat software for Macintosh System 7.0 or later.
	ACROUNIX	Contains the Acrobat software for UNIX platforms: HP-UX, SunOS and Solaris systems.
	ACROWIN	Contains the Acrobat software for the Windows 3.1, Windows 95, and Windows NT 3.5 or later systems.
CONSULT		This is *The ISDN Consultant* book. Open the ISDN.PDF file to get started.
FORMS		Here you will find the forms that were referenced in the book.
INDEX	INDEX	This directory contains the search indexes for the Acrobat documents on the CD-ROM.
netViz	disk1	The first installation disk of the netViz software.
	disk2	The second installation disk of the netViz software.

THE CD-ROM

TABLE D.1 CD-ROM directories. (Continued)

Level 2	Level 3	Purpose
NIUF		This directory contains the *A Catalog of National ISDN Solutions for Selected NIUF Applications* Acrobat files. Open the 0.PDF file (that's zero, not O) to get started.

■ INSTALLING THE SOFTWARE

Begin by installing the Adobe Acrobat reader to your system. You can access all of the documents from the CD-ROM or you can download the documents to your hard drive. If you download to the hard drive, you must recreate the directory structure to keep the search features operating.

Adobe Acrobat

■ *Macintosh Users*

Open the ISDN disk icon and double click on the ISDN-CD folder, and then the ACRO_V2 folder. Then open the ACROMAC folder and the DISK1 Folder, and double-click on the **ACOR-READ.MAC** icon.

The system requirements are:

- Macintosh 68020-68040: 2MB of application RAM
- Power Macintosh: 4MB of application RAM
- Apple System Software version 7.0 or later
- CD-ROM drive

■ *Windows Users*

Run **D:\ISDN-CD\ACRO_V2\ACROWIN\DISK1\ setup.exe** in the program. If D: is not your CD-ROM drive, then change D: to the drive letter for your CD-ROM drive.

The system requirements are:

- 386, 486, Pentium, or Pro processor-based computer
- Microsoft Windows 3.1, Windows 95, or Windows NT 3.5 or later
- 4MB of RAM
- CD-ROM drive

■ DOS users

Change to the CD-ROM drive by entering the drive letter at the command prompt. If your CD_ROM drive is D:, enter that at the prompt. The enter cd \ISDN-CD\ACRO_V1\ACRODOS and press enter. Now type in **install** and press enter.

The system requirements are:

- 386 or 486 processor-based computer
- DOS version 3.3 or greater
- 2MB of application RAM (4MB recommended)
- 5MB of hard disk space
- CD-ROM drive

■ UNIX Users

The installation instructions for each of the support platforms are located on the CD-ROM in the following locations:

For SunOS, Solaris, and HP-UX:

/cdrom/ISDN/ISDN-CD/ACRO_V2/ACROUNIX/instguid.txt

For Silicon Graphics:

/cdrom/ISDN/ISDN-CD/ACRO_V1/ACROSGI/readme.txt

netViz

The netViz 2.5 demonstration software can be installed by running the D:\ISDN-CD\netViz\disk1\setup.exe file from Windows Program Manager or the DOS command prompt.

■ Using Acrobat

The acrobat software is a simple, but powerful, software package that allows you to read any PDF file. In addition to the books included on this CD-ROM, you can use the software to read documents you receive in this file format.

To open a file, you use the File|Open command and follow the prompts to locate the document you want to read. Once the file is open you can scroll through the file, search it by any text or phrase, or search a group of documents using the index file to find the information you need. Figure D.1 is the main screen for the Acrobat Reader. This is the online guide for using the software, and is located in the help subdirectory where the Acrobat software was installed. Please take a little time to review the guide to learn how to unlock the powerful features in this software.

Some quick features to get to know: the VCR controls will help you page forward, backward, and to the beginning and end of the document you have open. On the left side of the screen is an optional display to quickly move you to different points in the document itself. You will also find certain areas of *The ISDN Consultant* that have red boxes, which are links to either the next chapter of the book or a related document. Anywhere you see a Internet URL address, like **http://ourworld.compuserve.com/homepages/roblee/**, if you have an Internet browser and a connection to the Internet, you can click on the URL and jump to that address.

On a final note, you can search the Acrobat files by using the find feature. This is a limited search capability that works only within the document you have open. You will find that *The ISDN Consultant* is hyperlinked so you can move between chapters and the table of contents and index. The ISDN.PDF file is your starting point to access each part of the book. The NIUF catalog has been hyperlinked also, but not to the same level. Included on the CD-ROM are the index files to enable searches across files. If you want to make use of these search capabilities, you will need to upgrade Acrobat Reader software to Acrobat Exchange. Information on this upgrade is available in the Adobe Acrobat help file.

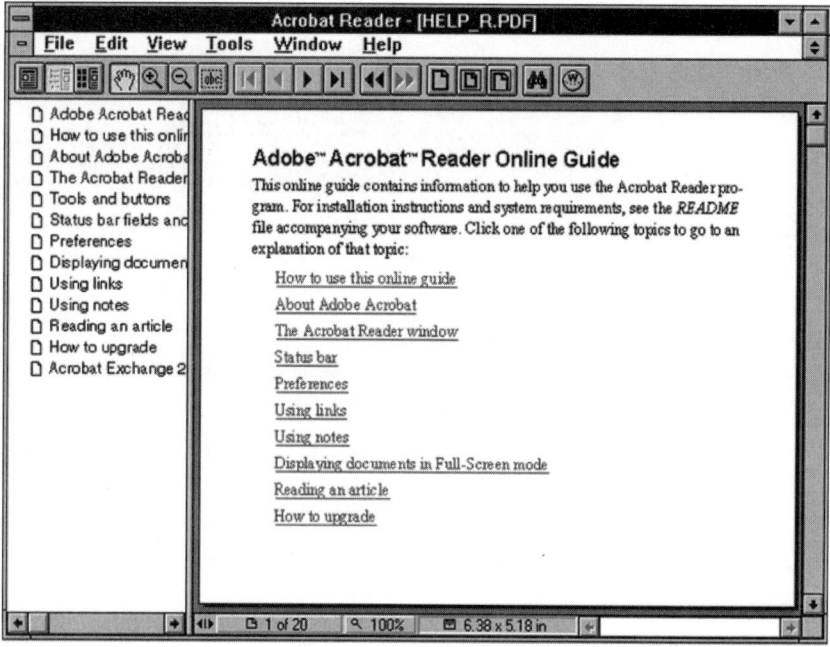

FIGURE D.1 ADOBE ACROBAT READER SCREEN AND HELP FILE.

Index

Numerics

110 Block 130
16550 UART chip 58
30-60-90 day review 107
800 Service 119, 123
900 Service 119, 123

A

Abbreviated Ringing and Delayed
    Ringing 68
Adobe Acrobat
    DOS Installation 226
    Internet Links 227
    Macintosh Installation 225
    Quick Start 227
    UNIX Installation 226
    Windows Installation 225

ADSL 16
Aids
    Definition box xxi
    Example box xxi
    Icons xxii
Alliance for Telecommunications
    Industry Solutions 6
Alpha Test 106
Ameritech 193
Asynchronous 42
AT&T 179, 206
    Custom ISDN 35
    ISDN tariffs 206
ATIS 33
ATM 41, 42
Audix 141
Automatic callback 68

B

B Channel 25
Backboards....................................51
Bandwidth
    Comparison chart.....................13
    Definition 15
Bearer channel 20
Bell Atlantic................................194
BellSouth 195
Beta review 107
Beta test......................................107
Beta test plan..............................107
Books
    ISDN Technology 219
    Network & Systems Design...220
    People and Project Skills 220
BRI
    2B+D 24
    Multiple devices......................26
    Voice services table 67
Bridge...59
Bright Idea
    ATM...43
    Centrex...................................121
    Circuit consolidation to PRI 29
    Dynamic line allocation 119
    LAN ISDN sharing 59
    Network design packages 100
    Small office phone solution 65
    Telecommuting 70
    Water Districts, Schools,
        Ecology 176
Broadband ISDN.....................28, 41
Broadcast audio...........................218
Broadcast industry 144
BT 179

C

Cable modems...............................12
Cable TV.....................................172
CACH EKTS 68
California ISDN User Group 214
Call Appearance Call Handling 68
Call Appearance Call Handling
    Electronic Key Telephone
    Service 68
Call appearances 132
Call Bridging................................68
Call Forwarding 67
Call Forwarding Don't Answer 67
Call Forwarding Interface Busy.....67
Calling Number Identification 68
Capability packages 70
Carrier 33
Caution
    Cable modems........................15
    D Channel backup...................27
    Internal adapters....................152
    ISDN power 28, 57
    Mixing local and long distance
        services......................122
    Passive bus..............................53
    Public packet networks 75
    Reusing existing wiring 51
    U Interface only devices 54
    X.25...26
CD-ROM
    Directories............................224
    Equipment resources...............94

Building wire closet layout diagram 50

INDEX

Hardware & software
    resources 185
ISDN voice features 67
netViz 2.5 demonstration 92
Order checklist & form 73
Problem Definition
    Worksheet 86
Review/Bid template 106
Standards 34
Cell 42
Central office
    Definition 10
    Line card 23
Centrex 120
Cincinnati Bell 197
CO 21
CODEC 169
Computer systems 57
Consultants 217

D

D Channel 25
    Backup 27
    BRI 25
    PRI 27
Data channel 20
Datacommunications primer 148
Dedicated voice services 118
Design Question List 97
Desktop videoconferencing 175
DID 120
Digi International Datafire card 65
Digital Subscriber Line
    Background 16
    HDSL 16
    VDSL 16

Directory Number 64
Directory Number Appearance 67
Distance learning
    videoconference 174
DMA 153
DN 64
Drop and Insert 142
    Example 143
    Functionality 56
DS-1 27
DS-2 28
DS-3 28

E

E0 28
E1 28
E2 28
E3 28
EKTS 36, 132
E-mail address 183
Encapsulated 40
Equipment 93
Euro-ISDN 11

F

Feature keys 132
Florida ISDN User Group 214
Frame relay 37
Framing bits 28
France Telecom tariff example 210
Frozen specifications 106

G

GTE 198

H

Hold/Retrieve 68
Home office with data and key
    system applications 135
Home wiring diagram 125
Home wiring reuse diagram 128
Hunt group 121

I

I/O 150
IBM 7845 NT1 57
Idea generation 88
Identifying the missing links 95
Implementation plan 107
Inbound 118
Indonesia ISDN & IN User
    Forum 215
Intel ProShare 66
Intel web site 66
Intercom 68
Interface points 52
    diagram 53
Internal adapters 152
Inverse multiplexor 56
IRQ 152
ISDN 3
    Availability, U.S. 10
    Benefits for dedicated telephone
        services 123
    BRI rate 5
    Centrex 177
    Channels 24
    Compression 149
    Data circuit features table 70
    Data features 69

Definition 20
E1 rate 5
Home office 134
In the home 124
Internal adapters 150
Messaging diagram 31
nB+D 20
Network connection 150
OSI Model 40
PBX based offices 136
PCS integration 38
PRI rate 5
Rate chart 20
Serial port 150
Small office 129
Synchronous interface 150
Voice Call Messaging 30
Voice features 67
ISDN design issues checklist 95
ISDN equipment 55
ITU 32
    CCITT 32
    International Telecommunications
        Union 6

K

Key system 35, 55

L

LAN routers 59
LCD 61
LD 21
LDDS 206
Local loop 23
Long distance carrier contacts 206

Long distance companies............205

M

Macintosh......................................58
Manufacturing application
    Final Design..............................99
    Physical design diagram 91
    Problem definition 87
    Process flow diagram...............90
MCI...206
Message Waiting Indicator 68
Messaging 25, 30
Midwinter Centrex videoconference
    design 177
MODEM 148
Modem compression....................149
Modem internetworking 38
MPEG-2 169
Multicasting 38
Multiple virtual lines......................68
Multipoint video/data
    conferencing..........................37
Multi-rate calls 28

N

Narrow the solutions 89
National ISDN 34
National ISDN-1 34
    Availability 10
National ISDN-2 ....35, 131
    Availability 10
    Features 36
National ISDN-3 37
    Features 37
netViz...226

netViz 2.5.....................................90
Network hubs................................60
Network termination devices 54
Network terminators 52
Nevada Bell.................................201
New York ISDN User Group.......214
NIUF web page.............................71
Non-U.S. local carrier contacts....207
North American ISDN Users'
    Forum..................................215
NPA 66
NT 52
NT1 52
NT2 52
NYNEX 201

O

Office connection examples.........132
On-line resource sites..................215
On-line services providers 219
Order checklist..............................74
Order codes table 72
Order process 71
OSI...38
    Diagram 39
    Model....................................38
Outbound 118

P

Pacific Bell.................................202
Pacific Region ISDN/Data User
    Forum..................................215
Packet data 25
Passive bus...................................54
Patch cords...................................49

Patch panel
    Concept diagram 50
    Definition 49
PBX 35, 54, 55, 121
    BRI 142
    Remote system links 141
PCS 37
People skills 94
Phone network 21
    Class 4 Switch 23
    Class 5 Switch 23
    Conceptual network diagram ... 22
    ISDN path 23
    Local call path 22
    Long distance call path 21
Physical design 97
POE 51
Port 55
PRI 26
    PBX 138
Private networks 123
problem definition 84
Problem definition template 86
Problem solving process diagram .. 85
Process diagram 83
Process flow 106
Project lifecycle 104, 105
Project tests 112
Protocols 38
    BONDED 151
    H.320 .... 40, 168
    MPPP 151
    V.110 .... 40, 150
    V.120 .... 40, 150
Public utility commissions 76
PUC 76

Punch down block 49
    Concept diagram 50
PVC 37

Q

Quyen Systems, Inc. 90

R

Redirecting number delivery 68
Regulatory process 78
Repeater device 24
Request for quotation 101
Resellers 217
Resources
    Data networks 186
    NT1 185
    Telephone suppliers 186
    Videoconferencing 188
Review & bid outline 102
RFQ 101
Rollout 107
Router 59

S

S/T 53
Same-to-different 151
Same-to-same 151
Sample application
    BRI vs. PRI tariffs 143
    Telephone line load
        balancing 119
Sample implementation plan 108
SAP 66
Satellite broadcasts 172
SBC 203

INDEX

Service Address Point Identifier 66
Service Profile IDentifier 64
Signaling System 7 ....21, 75
Six Party Conference Calling 68
SMDS 41
SNET 204
Software 217
Solution development 88
South African ISDN Forum 215
SPID 64
    Formats 66
Sprint 206
Standards
    General overview 6
Standards bodies 32
Standards creation flow 33
Support 113
Surfing, definition 17
Switch DS1 36

T

T interface 52
T-1 ....27, 55
Tandem 21
Tariffs 76
    Definition 76
    Goods & services 79
    Pricing & commitments 80
Technology design 106
TEI 66
Telephone systems 55
Telephones 56
    Dialing process 6
    Digital conversion 4
    Human voice 4
Telstra tariff example 209

Terminal adapters 25, 57
Terminal Endpoint Identifier 66
Testing procedures 110
Texas ISDN User Group 215
Three-Way Conference Calling 68
TIA 32
    Telecommunications Industry
        Association 6
Transfer 68
Trunk 55

U

U Interface 52
U.S. telephone companies 190
    Contacts 192
    Territories 11
UPS 30, 131
US West 205
User Story
    Broadcast audio 144
    Call center 141
    Dana Point videoconference ..176
    Distance learning -
        music lessons 144
    High tech manufacturer 110
    Home office 134
    Hudson Foods 179
    Intermittent failure 114
    Recycling center 140
    Remote systems links 141
Users Groups 214
UTP 48

V

Video telephones 179

Videoconferencing
    Conference room
        systems 60, 172
    Desktop systems 61
    Elements and configuration ... 170
    Multimedia conferencing 62
    Picture phones 61
    Primer 168
    Protocols 40
    Systems 60
Voice mail 141

W

WATS 119
Whiteboard 175
Wire cabinet 51

Wire closet
    Definition 47
    Designs 49
    Small office configuration 130
Wiring
    Categories 46
    Category 3 46
    Category 5 46
    Category Table 47
    Constraints 93
    Home based office 51
    Overview 45
    Reusing wire 51
    RJ connector diagram 48
    Shielded vs. Unshielded 47

X

X.25 25

LICENSE AGREEMENT AND LIMITED WARRANTY

READ THE FOLLOWING TERMS AND CONDITIONS CAREFULLY BEFORE OPENING THIS SOFTWARE MEDIA PACKAGE. THIS LEGAL DOCUMENT IS AN AGREEMENT BETWEEN YOU AND PRENTICE-HALL, INC. (THE "COMPANY"). BY OPENING THIS SEALED SOFTWARE MEDIA PACKAGE, YOU ARE AGREEING TO BE BOUND BY THESE TERMS AND CONDITIONS. IF YOU DO NOT AGREE WITH THESE TERMS AND CONDITIONS, DO NOT OPEN THE SOFTWARE MEDIA PACKAGE. PROMPTLY RETURN THE UNOPENED SOFTWARE MEDIA PACKAGE AND ALL ACCOMPANYING ITEMS TO THE PLACE YOU OBTAINED THEM FOR A FULL REFUND OF ANY SUMS YOU HAVE PAID.

1. **GRANT OF LICENSE:** In consideration of your payment of the license fee, which is part of the price you paid for this product, and your agreement to abide by the terms and conditions of this Agreement, the Company grants to you a nonexclusive right to use and display the copy of the enclosed software program (hereinafter the "SOFTWARE") on a single computer (i.e., with a single CPU) at a single location so long as you comply with the terms of this Agreement. The Company reserves all rights not expressly granted to you under this Agreement.

2. **OWNERSHIP OF SOFTWARE:** You own only the magnetic or physical media (the enclosed SOFTWARE) on which the SOFTWARE is recorded or fixed, but the Company retains all the rights, title, and ownership to the SOFTWARE recorded on the original SOFTWARE copy(ies) and all subsequent copies of the SOFTWARE, regardless of the form or media on which the original or other copies may exist. This license is not a sale of the original SOFTWARE or any copy to you.

3. **COPY RESTRICTIONS:** This SOFTWARE and the accompanying printed materials and user manual (the "Documentation") are the subject of copyright. You may not copy the Documentation or the SOFTWARE, except that you may make a single copy of the SOFTWARE for backup or archival purposes only. You may be held legally responsible for any copying or copyright infringement which is caused or encouraged by your failure to abide by the terms of this restriction.

4. **USE RESTRICTIONS:** You may not network the SOFTWARE or otherwise use it on more than one computer or computer terminal at the same time. You may physically transfer the SOFTWARE from one computer to another provided that the SOFTWARE is used on only one computer at a time. You may not distribute copies of the SOFTWARE or Documentation to others. You may not reverse engineer, disassemble, decompile, modify, adapt, translate, or create derivative works based on the SOFTWARE or the Documentation without the prior written consent of the Company.

5. **TRANSFER RESTRICTIONS:** The enclosed SOFTWARE is licensed only to you and may not be transferred to any one else without the prior written consent of the Company. Any unauthorized transfer of the SOFTWARE shall result in the immediate termination of this Agreement.

6. **TERMINATION:** This license is effective until terminated. This license will terminate automatically without notice from the Company and become null and void if you fail to comply with any provisions or limitations of this license. Upon termination, you shall destroy the Documentation and all copies of the SOFTWARE. All provisions of this Agreement as to warranties, limitation of liability, remedies or damages, and our ownership rights shall survive termination.

7. **MISCELLANEOUS:** This Agreement shall be construed in accordance with the laws of the United States of America and the State of New York and shall benefit the Company, its affiliates, and assignees.

8. **LIMITED WARRANTY AND DISCLAIMER OF WARRANTY:** The Company warrants that the SOFTWARE, when properly used in accordance with the Documentation, will operate in substantial conformity with the description of the SOFTWARE set forth in the Documentation. The Company does not warrant that the SOFTWARE will meet your requirements or that the operation of the SOFTWARE will be uninterrupted or error-free. The Company warrants that the

media on which the SOFTWARE is delivered shall be free from defects in materials and workmanship under normal use for a period of thirty (30) days from the date of your purchase. Your only remedy and the Company's only obligation under these limited warranties is, at the Company's option, return of the warranted item for a refund of any amounts paid by you or replacement of the item. Any replacement of SOFTWARE or media under the warranties shall not extend the original warranty period. The limited warranty set forth above shall not apply to any SOFTWARE which the Company determines in good faith has been subject to misuse, neglect, improper installation, repair, alteration, or damage by you. EXCEPT FOR THE EXPRESSED WARRANTIES SET FORTH ABOVE, THE COMPANY DISCLAIMS ALL WARRANTIES, EXPRESS OR IMPLIED, INCLUDING WITHOUT LIMITATION, THE IMPLIED WARRANTIES OF MERCHANTABILITY AND FITNESS FOR A PARTICULAR PURPOSE. EXCEPT FOR THE EXPRESS WARRANTY SET FORTH ABOVE, THE COMPANY DOES NOT WARRANT, GUARANTEE, OR MAKE ANY REPRESENTATION REGARDING THE USE OR THE RESULTS OF THE USE OF THE SOFTWARE IN TERMS OF ITS CORRECTNESS, ACCURACY, RELIABILITY, CURRENTNESS, OR OTHERWISE.

IN NO EVENT, SHALL THE COMPANY OR ITS EMPLOYEES, AGENTS, SUPPLIERS, OR CONTRACTORS BE LIABLE FOR ANY INCIDENTAL, INDIRECT, SPECIAL, OR CONSEQUENTIAL DAMAGES ARISING OUT OF OR IN CONNECTION WITH THE LICENSE GRANTED UNDER THIS AGREEMENT, OR FOR LOSS OF USE, LOSS OF DATA, LOSS OF INCOME OR PROFIT, OR OTHER LOSSES, SUSTAINED AS A RESULT OF INJURY TO ANY PERSON, OR LOSS OF OR DAMAGE TO PROPERTY, OR CLAIMS OF THIRD PARTIES, EVEN IF THE COMPANY OR AN AUTHORIZED REPRESENTATIVE OF THE COMPANY HAS BEEN ADVISED OF THE POSSIBILITY OF SUCH DAMAGES. IN NO EVENT SHALL LIABILITY OF THE COMPANY FOR DAMAGES WITH RESPECT TO THE SOFTWARE EXCEED THE AMOUNTS ACTUALLY PAID BY YOU, IF ANY, FOR THE SOFTWARE.

SOME JURISDICTIONS DO NOT ALLOW THE LIMITATION OF IMPLIED WARRANTIES OR LIABILITY FOR INCIDENTAL, INDIRECT, SPECIAL, OR CONSEQUENTIAL DAMAGES, SO THE ABOVE LIMITATIONS MAY NOT ALWAYS APPLY. THE WARRANTIES IN THIS AGREEMENT GIVE YOU SPECIFIC LEGAL RIGHTS AND YOU MAY ALSO HAVE OTHER RIGHTS WHICH VARY IN ACCORDANCE WITH LOCAL LAW.

ACKNOWLEDGMENT

YOU ACKNOWLEDGE THAT YOU HAVE READ THIS AGREEMENT, UNDERSTAND IT, AND AGREE TO BE BOUND BY ITS TERMS AND CONDITIONS. YOU ALSO AGREE THAT THIS AGREEMENT IS THE COMPLETE AND EXCLUSIVE STATEMENT OF THE AGREEMENT BETWEEN YOU AND THE COMPANY AND SUPERSEDES ALL PROPOSALS OR PRIOR AGREEMENTS, ORAL, OR WRITTEN, AND ANY OTHER COMMUNICATIONS BETWEEN YOU AND THE COMPANY OR ANY REPRESENTATIVE OF THE COMPANY RELATING TO THE SUBJECT MATTER OF THIS AGREEMENT.

Should you have any questions concerning this Agreement or if you wish to contact the Company for any reason, please contact in writing at the address below.

Robin Short
Prentice Hall PTR
One Lake Street
Upper Saddle River, New Jersey 07458